COSMOS CLOSE-UP

COSMOS CLOSE-UP

GILES SPARROW

Quercus

CONTENTS

INTRODUCTION

Zooming in on the Universe

For nearly all of human history, our observations and understanding of the world and the Universe around us have been limited by the physical abilities of the human body, and our eyesight in particular. Our eyes, which evolved over millions of years as life-saving sensory organs capable of alerting us to predators and potential prey amid the forests and plains of Africa, are nevertheless limited as scientific instruments – we can see a narrow range of light wavelengths, resolve a limited level of detail in small or distant objects and only detect objects at all if they are above a certain level of brightness. Until around 500 years ago, the only technological aids to our eyesight were crude spectacle lenses, capable of compensating for problems in the focus of our eyesight, but not much more. However, in the early 1600s, that began to change, with the discovery of the telescope.

Usually attributed to Dutch lens-maker Hans Lippershey, the principle of the first telescopes was simple: two convex lenses fixed at either end of a tube. The 'objective' lens at the front collected the more-or-less parallel rays of light coming from distant objects, and bent them onto converging paths, coming to a focus inside the telescope tube. The rear 'eyepiece' lens, meanwhile, intercepted the rays as they diverged again from the focus, bending their paths again so that, when they passed into the observer's eye, they appeared to be coming from a closer object. As a result, the image from a telescope appeared larger, and revealed more detail.

But a telescope does more than just magnify – it also intensifies images, since its light-collecting surface may be many times larger than the small pupil through which a human eye gathers light. It was this realization that inspired

early astronomers, including the great Italian physicist Galileo Galilei, to develop their own telescopes and turn them towards the heavens. Here, Galileo and his contemporaries discovered a plethora of previously unexpected features, ranging from satellites orbiting Jupiter to the changing phases of Venus, and from mountains on the Moon to countless individual stars, clusters and knots of gas embedded in the Milky Way.

The arrival of telescopic astronomy caused a scientific revolution that finally overthrew centuries of received wisdom. Most importantly of all, it began the long process of reassessing Earth's place in the Universe, relegating it from the centre of the Universe to just one of several worlds orbiting the Sun. Further discoveries in the centuries since then have seen our planet slip further from cosmic pre-eminence, until today we know that Earth is an unremarkable rocky planet orbiting an insignificant

star some 26,000 light years from the centre of the Milky Way galaxy. The Sun in turn is one of about 200 billion stars in our galaxy, and there are probably as many galaxies in the Universe as there are stars in the Milky Way.

Telescopes are our greatest tool for studying the distant Universe in detail. Today, large research instruments (such as those used to produce many of the images in this book) rely on mirrors rather than lenses – light entering the telescope is reflected onto converging paths as it bounces off a large 'primary' mirror whose surface is curved with mathematical precision. From here, the light is redirected from a secondary mirror towards either a camera or a variety of other scientific instruments, most of which rely on ultra-sensitive versions of the electronic imaging CCDs or 'charge-coupled devices' found

Using mirrors to collect the light (an innovation first introduced by James Gregory and Isaac Newton in the 1660s) avoids a range of problems found in lens-based telescopes, ranging from absorption of light in the glass, to various distortions or 'aberrations' produced in the final image.

The world's major observatories are mostly built on a few mountaintop sites, including Cerro Paranal in Chile, Mauna Kea in Hawaii and La Palma in the Canary Isles. Here, they lie above the majority of Earth's atmosphere and benefit from the clearest possible view of the night sky, situated above the majority of clouds, and with absorption of light and turbulence introduced by the atmospheric gases above kept to a minimum. The largest telescopes of all use multiple-mirror designs, with their primary reflector composed of several smaller interlocking hexagons. Made from relatively thin and slightly flexible materials, the mirrors are kept in alignment by computerized motors called 'actuators' that keep their overall curve precise as the telescope swings in various directions. Advanced 'adaptive optics' systems can even take the turbulence introduced in the upper atmosphere into account, adjusting the final image to 'filter out' the worst distortion.

The largest multiple-mirror telescopes currently in operation have collecting surfaces up to 10.4 m (34 ft) in diameter, and even larger instruments, up to 30 m (100 ft) in diameter are at various stages of construction and planning. What's more, observatories such as the Keck on Mauna Kea and the Very Large Telescope in Chile can combine the observations of multiple large telescopes using a technique called interferometry, allowing them to simulate the 'resolving power' of a single instrument several dozen metres across. Optical interferometry (developed from a technique invented for radio astronomy) can resolve fine details with angular sizes around 1 'milliseconds of arc' (roughly one two-millionth the diameter of the Full Moon).

Using these techniques, ground-based observatories can overcome the inevitable blurring introduced by the atmosphere, and today they can rival the sharpness of images from the most famous observatory of all, the orbiting Hubble Space Telescope. Launched in 1990, Hubble's 2.4-m (8-ft) mirror is relatively modest by modern standards, but its location outside of Earth's atmosphere gives it a unique clarity and allows it to detect the feeble light of the faintest and most distant objects.

Of course, Hubble is not alone in orbit – it is just the best known among dozens of telescopes that have been put into space over the past five decades. Most of these observatories, however, focus on radiations beyond visible light. Such radiations (similar in nature to normal light, but with more or less energy) range from high-energy gamma rays, X-rays and ultraviolet light, through the visible spectrum, down to the lower-energy infrared, microwaves and radio waves, with the lowest energy of all. They are emitted by a huge variety of objects, from violent exploding stars and stellar remnants in the case of the higher-energy, short-wavelength radiations, down to cool interstellar dust and gas in the case of the longer wavelengths. Earth's atmosphere blocks out most types of non-visible radiation (except for some infrared that can reach mountaintop observatories, and certain radio waves), and so the only way we can learn about a huge range of celestial phenomena is to observe them from space.

With an array of high-tech tools at their disposal, ranging from mountaintop observatories to orbiting telescopes (not to mention the robot space probes that have now ventured out across our solar system to study all the major planets at close proximity), modern astronomers can study the Universe in unprecedented detail, zooming in to study fine structures, distant objects and radiations beyond visible light, as revealed by the stunning images throughout this book.

By looking closer, they have also come to understand far more about the true nature of the Universe, and our special place within it. For despite all we have discovered, there is still one way in which planet Earth is unique – it is the only world we know of that has given rise not only to life, but also to intelligent minds with the curiosity to reach out across the cosmos with their technology, and attempt to grasp its meaning.

Measurement units and symbols used throughout the book:

Astronomical Unit (AU) – The average Earth–Sun distance of 149.6 million km (93 million miles).

Light year – The distance traveled by light in a vacuum during one Earth Year. One light year is equal to 9.5 trillion km, 5.9 million million miles, or 63,240 astronomical units.

◇ Distance to the object from Earth.
✕ Size of 'close-up' image.

THE SOLAR SYSTEM

The solar system is our immediate cosmic neighbourhood, containing an array of planets, moons and smaller objects dominated by our local star, the Sun. Objects in the solar system are close enough to Earth for our spacecraft to visit directly, and five decades of interplanetary exploration have produced some spectacular close-up images that have transformed our understanding of these nearby worlds.

The solar system is a region of space in which our local star, the Sun, exerts the dominant influence. By some estimates, the Sun's gravitational reach stretches out for almost 9.5 trillion km (5.9 million million miles) in every direction – so far out that light from the Sun takes an entire year to reach the deep-frozen chunks of ice making up the Oort Cloud – the most distant region of our solar system. Even if the region of solar influence is limited to the edge of the 'heliosphere' – where the stream of 'solar wind' particles flowing out from the Sun slows and falters against the pressure of stellar winds from countless other stars – the solar system still extends to around 100 times Earth's orbital diameter, encompassing all the major planets and most of the 'ice dwarf' worlds of the Kuiper Belt beyond Neptune.

The solar system also includes everything within that region – ranging from huge gas giant planets such as Jupiter, to smaller rocky worlds such as Earth itself, and many millions of individual meteoroids, asteroids and comets – chunks of rock and ice that fill the space between the major planets. Despite the presence of a huge number of individual objects, however, most of these are so small compared to the available volume of space that collisions are rare (though inevitable on long timescales).

Astronomers define a planet as an object in its own independent orbit around the Sun, which has sufficient gravity to pull itself into a spherical shape, and also exerts enough influence on its surroundings to 'clear out' smaller objects that attempt to share its orbit – either pulling them into orbit as satellites, or flinging them out into different paths around the Sun. By this definition, there are eight major planets, all travelling around the Sun in roughly circular orbits that lie in more or less the same flat plane around our star. Although this seems like the natural order of things, such well-ordered planetary orbits appear to be a rarity in the wider Universe.

Interplanetary distances are often measured in terms of 'astronomical units' or AU, where 1 AU is the average Earth–Sun distance, equivalent to 149.6 million km (93 million miles).

Close to the Sun lie four rocky planets – Mercury (the smallest, only 40 per cent larger than Earth's Moon), Venus (almost the same size as our own planet), Earth itself (the largest solid object in the solar system) and Mars (roughly half the size of Earth). Conditions across these four planets vary hugely, depending on their size (and hence their surface gravity) and their distance from the Sun. Mercury is a baking airless ball of rock, and Venus a choking 'hothouse' with a thick toxic atmosphere and a runaway greenhouse effect, while Mars is a cold, desert world with a thin atmosphere and extensive ice deposits.

Earth itself, in contrast, has a protective atmosphere of just the right density to create a warming blanket around our planet that in turn allows liquid water to exist in copious amounts on its surface (the only planet in the solar system where this is possible). Because water is a prerequisite for

INNER WORLDS

This artist's impression shows the rocky planets of the inner solar system in their orbits around the Sun (not to scale). The innermost planet, Mercury, completes an orbit in just 88 Earth days, following an elliptical path that takes it between 0.31 and 0.47 AU from the Sun. Venus, the second planet, has a more circular 225-day orbit at a distance of 0.72 AU. Next comes our homeworld, Earth itself, orbiting at an average distance of 1 AU in 365.26 days. Beyond Earth, the red planet Mars orbits the Sun in 1.88 Earth years. It also has a noticeably eccentric orbit, ranging between 1.4 and 1.7 AU from the Sun. The realm of the rocky planets is surrounded at its outer edge by the main asteroid belt, concentrated between 2.3 and 3.3 AU.

OUTER VIEW

This artist's impression focuses on the outer worlds of the solar system, beyond the asteroid belt that separates Mars and Jupiter, the innermost and largest giant. Jupiter orbits the Sun once every 11.9 Earth years at an average distance of 5.2 AU. Beyond it lies multi-ringed Saturn, orbiting in 29.5 years at a distance of 9.6 AU. Saturn is the most distant planet readily visible to the naked eye, and was the outermost planet known to ancient astronomers. Beyond it lies tilted Uranus, orbiting the Sun every 84 years at a distance of 19.2 AU (exactly twice that of Saturn). Deep blue Neptune is the outermost major planet, with an average distance of 30 AU from the Sun, and a 165-year orbit. The illustration also includes a representative Kuiper Belt object in the form of Pluto, which follows a highly eccentric 248-year orbit ranging between 30 and 49 AU from the Sun.

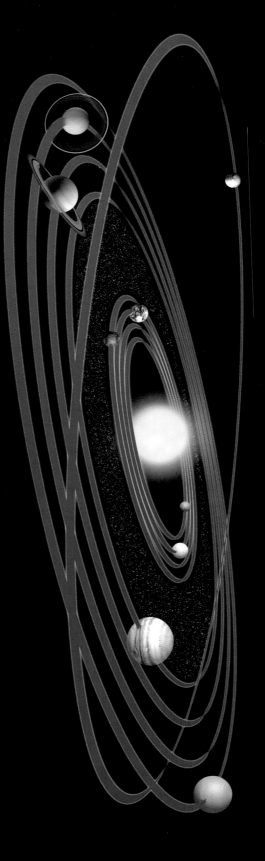

the evolution of life as we understand it, Earth is often said to lie in a 'Goldilocks zone' where the temperature is neither too hot, nor too cold, but just right. Earth's giant Moon, more than a quarter the size of our own planet, is also thought to play an important role in keeping Earth stable and protecting the development of life.

Beyond Mars lies a broad zone of rocky debris – the asteroid belt – and then comes the realm of the giant planets. In order from the Sun (and decreasing order of size), these are the colourful, turbulent Jupiter, the magnificently ringed Saturn, the enigmatic Uranus and the stormy Neptune. Uranus and Neptune are significantly smaller than their inner neighbours, and some astronomers classify them as 'ice giants' with slushy interiors made of chemicals such as water and ammonia, in contrast to the inner 'gas giants' dominated by gaseous and liquid hydrogen.

Each of the giant planets exerts so much gravity that it lies at the centre of its own complex satellite system, orbited by numerous true moons and countless smaller particles jostling one another within intricate ring systems (all the giant planets have rings of one sort or another, but those of Saturn are by far the most impressive).

The larger satellites of the outer planets are complex worlds in their own right, some of which rival Mercury in size. Despite their distance from the Sun and relatively small size, they often display unusual geological features and even surface activity such as volcanoes and geysers, thanks to the heat generated by tides as they orbit their parent planets.

Beyond the orbit of Neptune lie a host of icy worlds. The innermost and largest of these are the 'ice dwarfs' that orbit in the doughnut-shaped Kuiper Belt – some of which (such as Pluto and Eris) approach the size of the planet Mercury. Further out lies a spherical halo containing trillions of smaller chunks of ice – the Kuiper Belt. Each of these ice blocks is a potential comet nucleus that can turn into a spectacular comet if its orbit is disturbed and it comes closer to the Sun.

The solar system originated as a disc of debris that remained behind when the Sun formed around 5 billion years ago. Close in, heat from the young star evaporated chemicals with low melting points, and the gases were then driven away from the Sun by the fierce solar wind. The remaining material coalesced through random collisions until some small 'planetesimal' worlds developed enough gravity to start pulling in debris from their surroundings, rapidly growing into full-blown planets. Further out, gas and icy chemicals survived, and huge eddies in the 'protoplanetary' disc collapsed to form giant planets whose interiors gradually separated out into solid, liquid and gaseous layers. In among the planets, countless smaller worlds of rock and ice survived, many of which have since collided with the larger worlds, or interacted with them to be flung into new orbits.

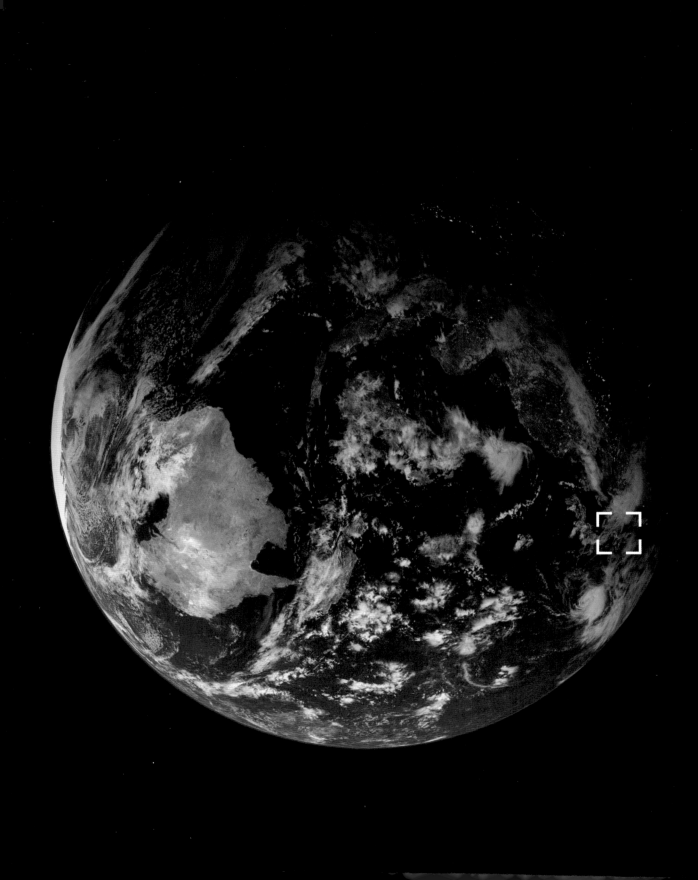

EARTH: OCEAN PLANET

Seen from space, our home planet is dominated by water – little wonder this detailed NASA satellite image of the globe was nicknamed the 'Blue Marble'. Water covers 71 per cent of Earth's surface, most of it collected in ocean basins that may be several kilometres deep. Liquid surface water is one of the things that makes Earth unique in the solar system – it's only possible because of our planet's perfect position in the 'Goldilocks zone' around the Sun (neither hot enough for water to boil, nor cold enough for it to freeze solid) and the blanketing effect of our substantial atmosphere.

■ WATERWORLD

An orbital view shows the southern Kuril Islands and Hokkaido, Japan, in the midst of the vast Pacific Ocean. Water plays a crucial role in shaping Earth's environment – it is constantly being transformed between liquid, solid and vapour forms in a complex 'water cycle'. Water evaporating into the atmosphere condenses into clouds and falls back down to Earth either as rain, which carves and shapes the landscape as it finds its way back to the sea, or as snow, which accumulates in glaciers and icecaps. Substantial amounts of water are also locked away in rocky minerals, and water is also vital to the evolution and sustenance of life.

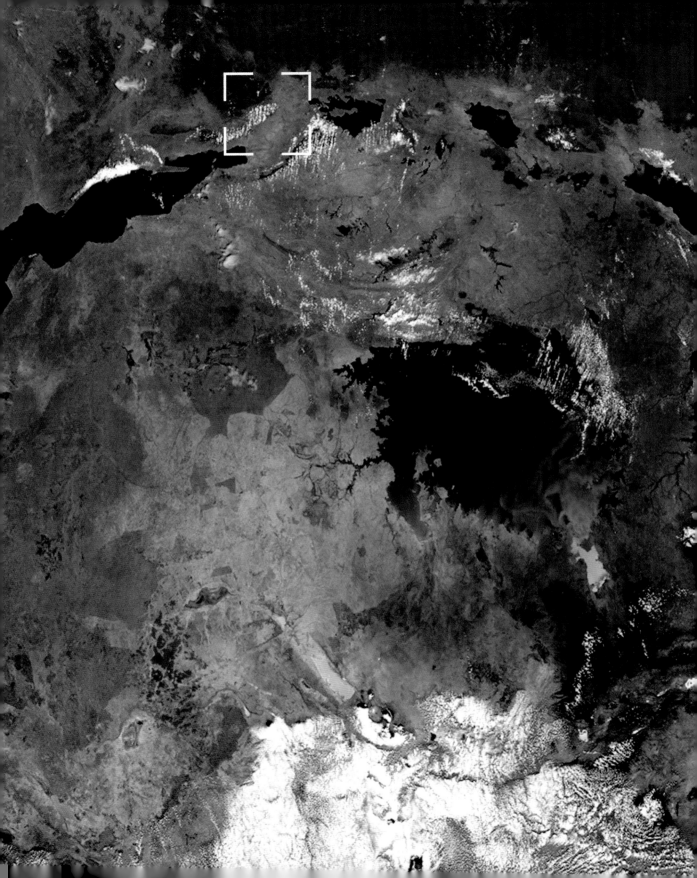

EARTH: TECTONIC CRUST

Another unique feature of the Earth is the nature of its geological activity. The process of formation generates a great deal of heat inside any planetary body, and the decay of radioactive elements deep within releases more energy. The larger a planet is, the longer it takes to cool down and solidify, and Earth's interior has remained hot and molten for more than 4.5 billion years. Heat escaping from the core escapes through a deep layer of semi-solid rock called the mantle, and friction between this churning layer and the thin outer crust causes the crust to fragment into mobile 'tectonic plates', which move across the surface at rates of a few centimetres per year.

■ RIFT VALLEY

The East African Rift Valley is one of the best places on Earth to see tectonics in action – here two tectonic plates are pulling apart and volcanic activity is slowly producing new crust in between them. Newly formed crust is typically much thinner than the continental crust (a few kilometres deep compared to tens of kilometres in the ancient 'cratons' at the hearts of the continents), and as a result it is rarely seen on land – most of Earth's younger crust lies on the ocean floors, spreading slowly outwards from volcanic deep-sea ridges.

EARTH: MOUNTAIN BUILDING

Where Earth's tectonic plates come together, the results can be spectacular, as one plate is forced to give way to another on a timescale of millions of years. If the collision is between oceanic and continental crust, the thin oceanic crust is usually forced downwards, melting as it plunges back into the mantle below, but releasing heat in the process that can trigger volcanoes along the edge of the overlying continent. This is how volcanic mountain chains such as the Andes of South America have formed.

■ EVEREST FROM ABOVE

Collisions between two blocks of continental crust can be even more impressive, as seen where the Indo-Australian plate has collided with Asia over the past 70 million years to produce the Himalayas. The collision started at a rate of about 15 cm (6 in) per year – high speed in tectonic terms – and has forced the southern edge of Eurasia over the top of the Indian subcontinent, creating a towering mountain range with the Tibetan Plateau behind it.

EARTH'S MOON: SEA OF TRANQUILLITY

The largest features on the surface of the Moon are dark lowland plains known as maria or 'seas'. They formed when volcanic eruptions flooded deep basins gouged out by huge impacts early in the Moon's history. This activity wiped whole areas of the Moon clear of the craters that had accumulated in the first half-billion years of its history, and as the lava solidified, it formed a new surface on which craters have continued to form, at a much reduced rate, ever since. The Mare Tranquillitatis or Sea of Tranquillity is one of the largest and best known of all lunar seas.

■ LUNAR REGOLITH

In 1969, the Sea of Tranquillity was the destination for the first manned lunar landing by the astronauts of Apollo 11. While the region appears smooth and clear from space, in reality it proved to be strewn with rubble and the Eagle lunar module touched down with barely 30 seconds of fuel to spare. The upper few metres of the mare are made from a jumble of rocks and dust of different sizes, known as the 'regolith'. This is the result of constant pounding by impacts of all sizes in the 3 billion or so years since they formed.

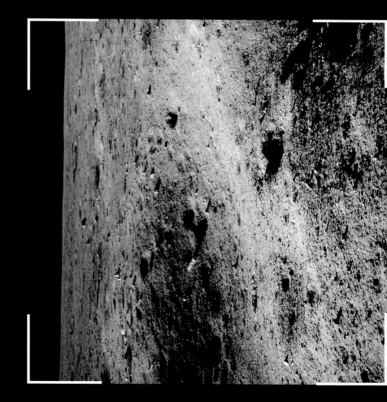

5

EARTH'S MOON: TSIOLKOVSKY CRATER

While the near side of the Moon has a broad mix of dark seas and bright, cratered highlands, the lunar far side, first photographed in 1959, is very different, dominated entirely by highlands. Over the 4.5 billion years since the Moon formed, tidal forces caused by the pull of Earth's gravity have slowed down our satellite's rotation so that it matches with its orbital period around the Earth. Both are now equivalent to 27.3 Earth days, so the Moon keeps one face permanently turned towards Earth. Astronomers think this effect also pulled the Moon's once-molten core closer to Earth, so that volcanic activity and escaping heat had a greater effect on the near side.

APOLLO'S VIEW

The crater Tsiolkovsky is one of the few obvious dark features on the Moon's far side. It was discovered by the Soviet probe Lunik 3 during its October 1959 flyby, and named after the Russian rocketry pioneer Konstantin Tsiolkovsky. Some 180 km (112 miles) in diameter, the crater is some 5 km (3 miles) deep, with a mountainous peak at its centre, surrounded by a range of volcanic features formed as lava erupted onto the surface. Astronomers still aren't sure quite why Tsiolkovsky saw so much more volcanic activity than the rest of the far side.

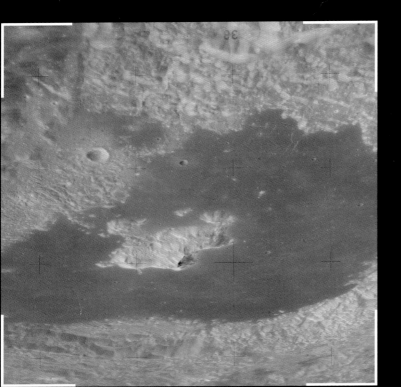

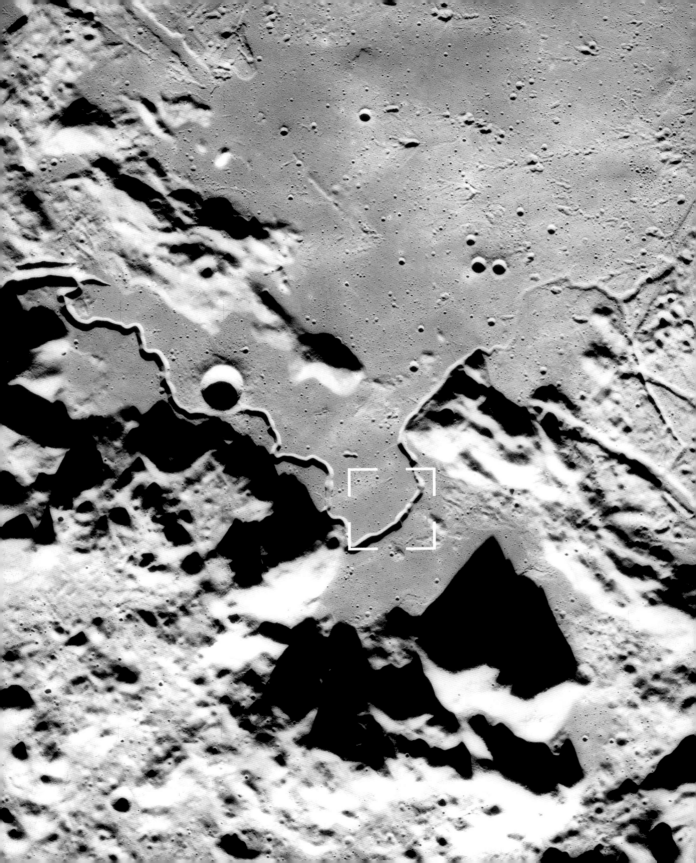

6

EARTH'S MOON: THE LUNAR APENNINES

The great curving chain of the Montes Apenninus or Lunar Apennine Mountains stretches for some 600 km (370 miles) across mid-northern latitudes of the lunar near side, marking the edge of the Mare Imbrium or Sea of Showers. While mountains on Earth are typically pushed up by tectonic forces from within the planet, lunar mountain ranges are typically the disguised and broken-down rims of huge ancient impact basins. The Apennines are no exception – they are formed from material pushed out during the formation of the Imbrium Basin around 3.9 billion years ago, and subsequently part-buried by eruptions.

■ IN THE SHADOW OF MONS HADLEY

At the northeast end of the Lunar Apennines stands Mons Hadley or Hadley Mountain, a towering triangular peak that formed the target for the Apollo 15 manned mission in 1971 (shown behind astronaut Jim Irwin and the Lunar Roving Vehicle in this image). Aside from the mountains themselves, the other major feature of this region is Hadley Rille, a 'sinuous rille' valley that winds its way across some 135 km (84 miles) of lunar lava plains and reaches up to 1 km (0.6 miles) across. The rille proved to be a collapsed lava tube – a channel formed where volcanic mare lava once flowed in an underground channel, whose roof later collapsed once the lava disappeared.

EARTH'S MOON: COPERNICUS CRATER

Among the most prominent of all lunar craters, Copernicus has a diameter of some 93 km (58 miles). It is surrounded by rays of bright 'ejecta' material that was pulverized and sprayed out across the lunar landscape when the crater formed some 800 million years ago. In places this ejecta can be traced for more than 800 km (500 miles) from the crater itself. Such 'ejecta blankets' provide an important way for scientists to date the age of different lunar features, and also offer an intriguing glimpse of the minerals from beneath the Moon's surface that are excavated by impacts.

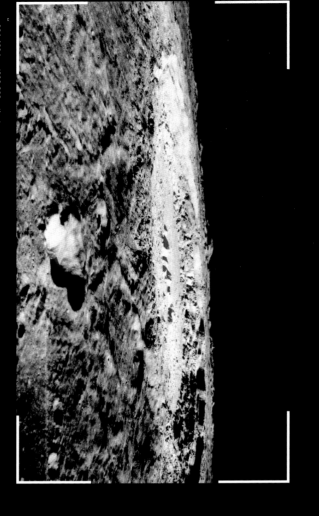

384,400 km (238,800 miles)

93 km (58 miles)

■ OVER THE EDGE

In the mid-1960s, NASA launched a number of robot space probes to investigate the Moon in preparation for the Apollo manned missions. Lunar Orbiter 2 snapped this oblique view across the crater in November 1966 – at the time, the media hailed it as one of the 'pictures of the century'. Collapsed terraces form a crater wall some 3.8 km (2.4 miles) deep, with central peaks rising up to 1.2 km (0.75 miles) high in the centre. While the vast majority of large craters formed in the first billion years or so of lunar history, Copernicus is much younger – around 800 million years old – explaining its relatively fresh appearance.

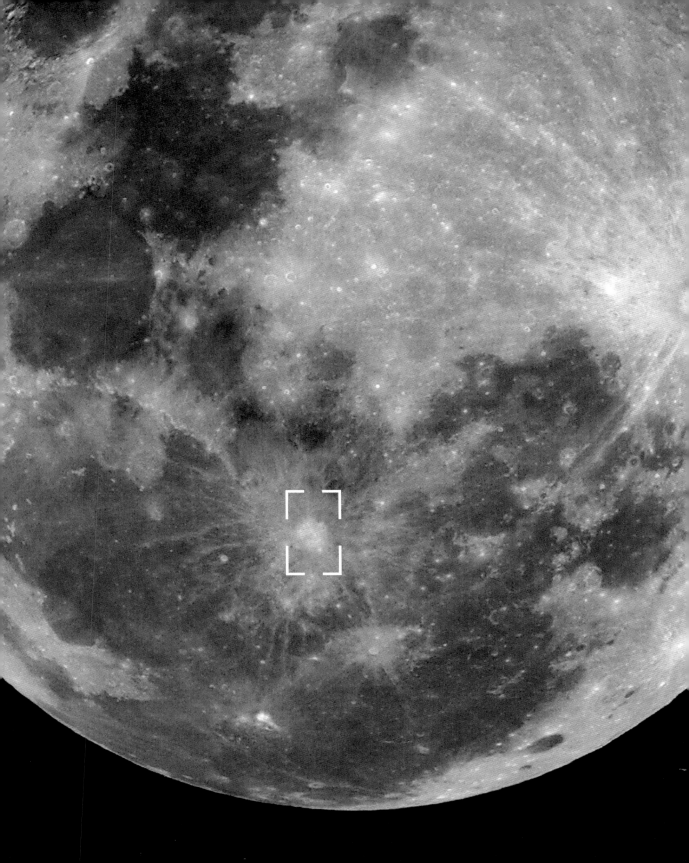

From Earth, our Sun appears as an unbearably bright ball of light in the sky, with a sharply defined outer edge called the photosphere. However, the reality is rather different – the Sun is nothing more or less than a huge ball of gas, shining due to nuclear reactions that take place at huge temperatures and pressures in its core. The photosphere is merely a region, 1,000 km (620 miles) or so deep, where the gas becomes transparent and light escapes outwards. While the Sun's visible disc has a diameter of about 1.4 million km (865,000 miles), its outer atmosphere, the corona, stretches out to more than three times this size.

■ TURBULENT SURFACE
Normally, the solar corona is a tenuous layer of gases heated to tremendous temperatures by the Sun's magnetic field – while the temperature of the photosphere is around 5,500°C (9,900°F), the corona can reach temperatures of 1–2 million °C (1.8–3.6 million °F). However, violent events close to the solar surface, where the complex magnetic field gets tangled together and rearranges itself, can sometimes release huge amounts of cooler, denser material. Propelled out across the solar system, these clouds form solar flares or coronal mass ejections that may contain more than a billion tonnes of matter.

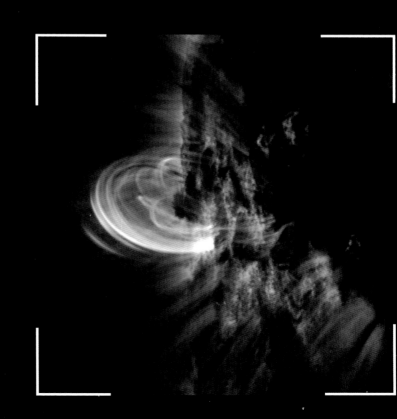

THE SUN: SURFACE ACTIVITY

This ultraviolet view of the Sun's surface from the TRACE (Transition Region And Coronal Explorer) satellite reveals the turbulent reality of the apparently placid Sun. Jets of hot gas escape along the tangled magnetic field where it emerges from the surface, forming loops called prominences. While magnetic fields in planets like Earth are semi-permanent and change only slowly, the Sun's magnetism is generated by moving masses of electrically charged gas, and waxes and wanes in an 11-year cycle, with a huge effect on solar activity.

■ FINE STRUCTURE

A stunning close-up from the Swedish Solar Telescope on La Palma in the Canary Isles, blocks out most of the Sun's radiation to reveal detail at the narrow 'hydrogen-alpha' wavelength. Here, the Sun's visible surface is resolved into flame-like pillars or 'spicules', each thousands of kilometres long, swirling in the solar magnetic field.

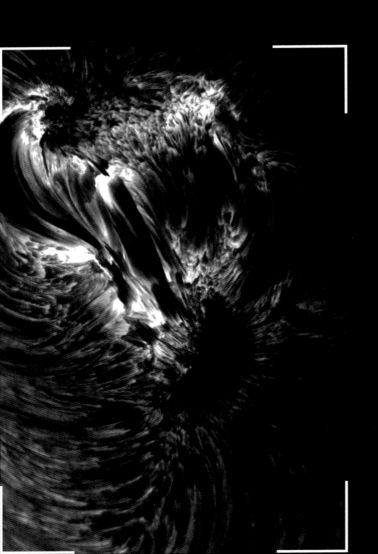

Sunspots are the most obvious solar features visible from Earth – dark patches in the photosphere that typically last for days or weeks, moving across the face of the Sun and revealing its complex 'differential' rotation (equatorial regions rotate every 25 days, while higher latitudes move more slowly, rotating in up to 34 days). Sunspots appear dark because they are cool compared to the surrounding photosphere – however, they still typically have temperatures of around 3,500°C (6,300°F). Their numbers increase and fall off, and they shift their general position on the solar disc, in line with the 11-year solar magnetic cycle.

■ SOLAR GRANULATION

Sunspots form where loops of the solar magnetic field force their way out through the photosphere, creating a cooler, lower-density region. In this stunning Swedish Solar Telescope image, the surrounding 'granulation' cells are thousands of kilometres wide. Each individual cell, with a bright centre and a dark edge, marks the top of a rising column of hot gas emerging from within the Sun. As it reaches the photosphere and becomes transparent, the heat trapped within it escapes as sunlight and other radiations. The gas itself cools and darkens before falling back to make way for more upwelling gas.

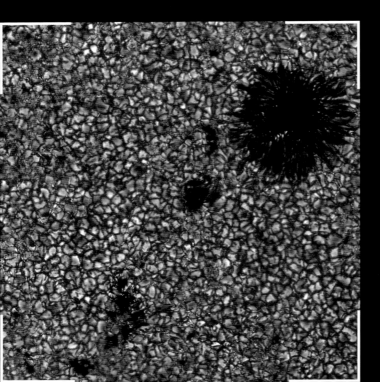

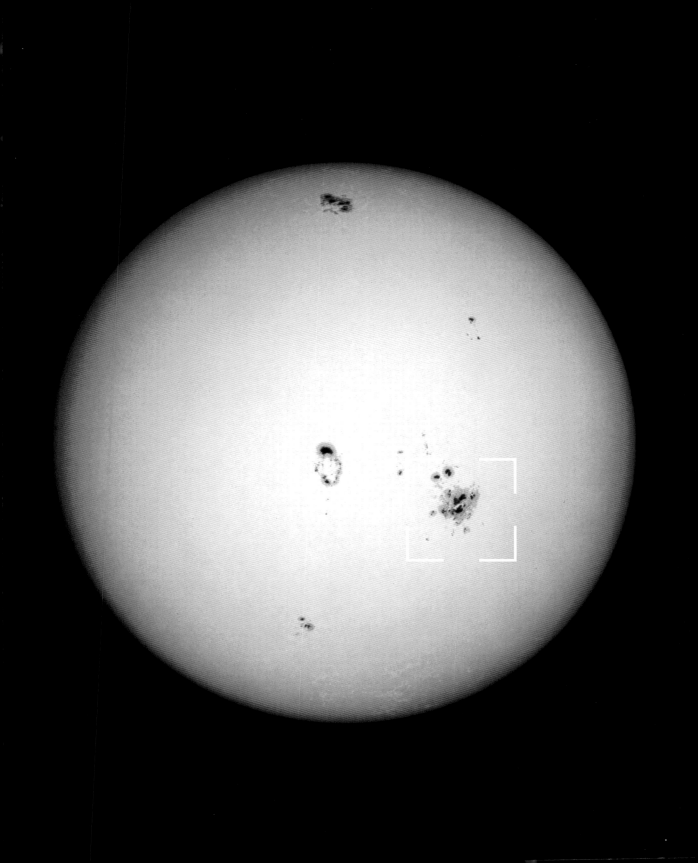

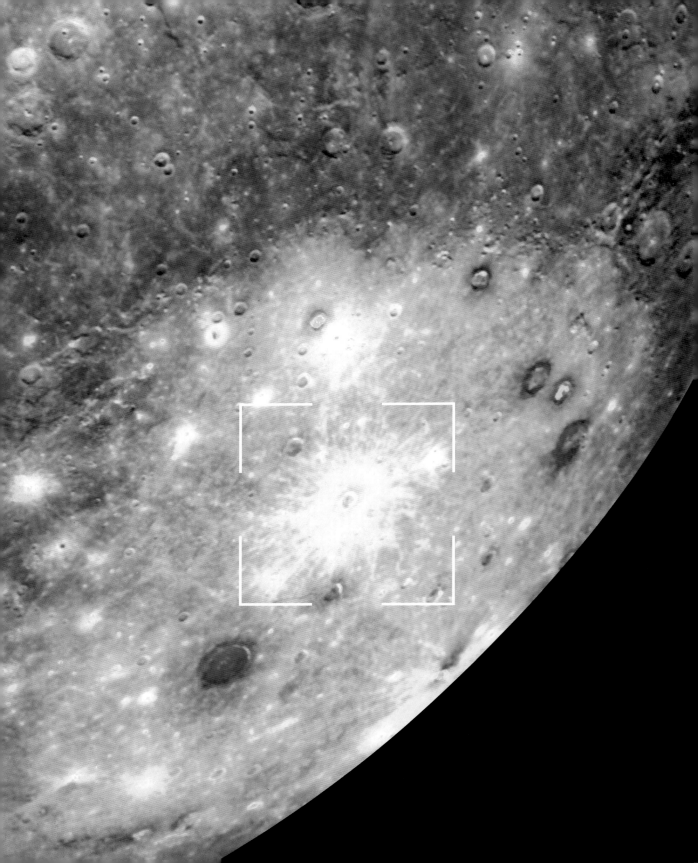

MERCURY: CALORIS BASIN

At first glance, Mercury seems remarkably similar to the Moon – a grey, airless world dominated by craters. As the smallest of the four terrestrial planets, Mercury cooled fairly rapidly, losing heat and slowing its geological activity several billion years ago. Since then, the Sun's tidal forces have slowed Mercury's rotation so that it rotates three times in every two 88-day orbits around the Sun. This unique arrangement means that most parts of Mercury only see a sunrise every two years, and creates an extreme range of temperatures, varying from –190°C (–342°F) to 430°C (806°F).

■ THE SPIDER

Foremost among Mercury's countless craters is the Caloris Basin. At 1,550 km (960 miles) across, it is the largest impact basin in the entire solar system, rimmed by a triple ring of mountains up to 2 km (1.2 miles) high. Despite the basin's size, its central regions are remarkably flat – mostly thanks to volcanic activity that resurfaced its interior shortly after its initial formation about 3.8 billion years ago. However, right at the centre lies this curious feature – an impact crater surrounded by radiating troughs that may mark the origin of at least some of the volcanic lava. Unsurprisingly, this strange crater is known as 'the spider'.

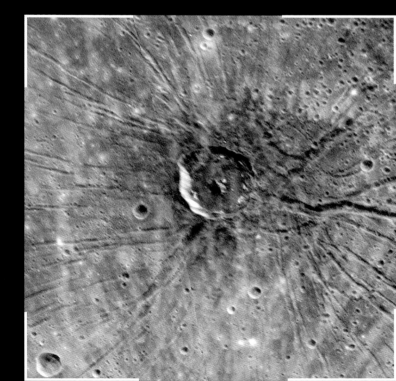

MERCURY: VOLCANIC ACTIVITY

Mercury's rapid 88-day orbit around the Sun makes it hard to reach using space probes from Earth – it moves 20 km/s (12 miles per second) faster through space than our own planet. Mariner 10 managed a series of three flybys in the early 1970s, but these left roughly half the planet unmapped, and it's only in the last few years that the MESSENGER probe has returned to the innermost planet, making its own series of flybys prior to entering orbit around Mercury in 2011. Among MESSENGER's first discoveries was a large basin called Rachmaninoff, roughly 290 km (180 miles) across.

■ BRIGHT PATCHES

The MESSENGER probe has travelled to Mercury equipped with a variety of modern instruments suitable for analysing minerals on the planet's surface. This enhanced-colour image of the Rachmaninoff region, for instance, reveals a bright patch of unusual yellow-white material centred on a small nearby depression – planetary scientists think this was probably a fissure that spewed volcanic material across the surrounding surface. Evidence from MESSENGER suggests that Mercury was volcanically active through much of its life, perhaps because its unusually large metal core remained molten for a long time.

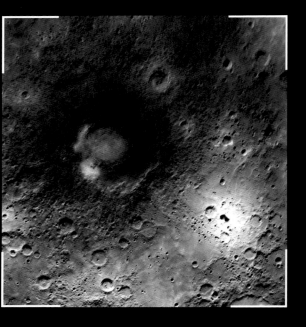

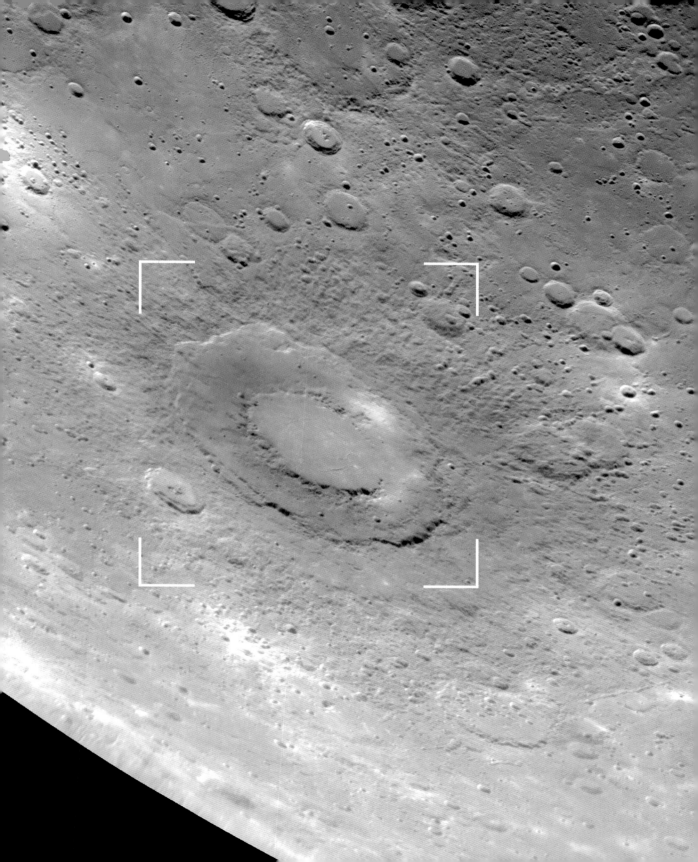

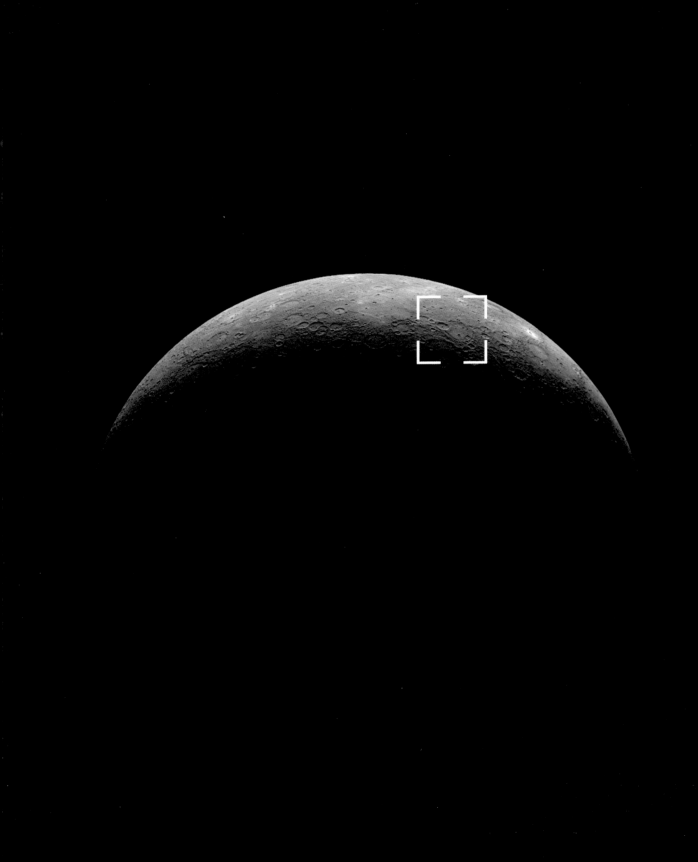

[13] MERCURY: SHRINKING PLANET

Astronomers have known for a while that Mercury's core of iron and nickel is unusually large compared to its overall size. This is almost certainly the result of a cataclysmic interplanetary collision between Mercury and another small 'protoplanet' early in its history – the collision stripped away much of Mercury's rocky mantle but left it with the core of a much larger planet. Because of this imbalance, the changing temperature of the core has had a major influence on the development of Mercury. Early in it history, the core heated and swelled up, causing the surface to expand, before later cooling, shrinking and collapsing inwards.

■ DISCOVERY RUPES

The signs of Mercury's unusual expansion and shrinkage are written across its landscape in the form of unusual escarpments known as rupes. Often running for hundreds of kilometres, these features separate different parts of Mercury's surface with cliffs towering up to 1 km (0.6 miles) high. Often the surface on either side of a rupes is unaltered, leaving basins and craters simply split in two. The rupes are the result of Mercury's surface expanding and splitting apart before the planet shrank and the now-outsized crust fell back, jamming itself together like badly fitted pieces of a jigsaw puzzle.

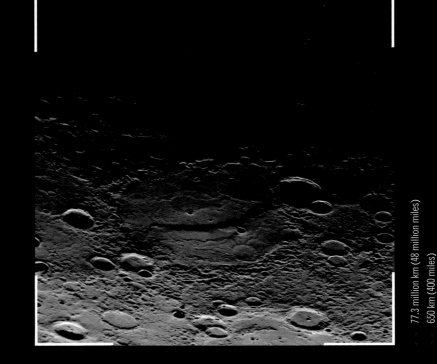

›› 77.3 million km (48 million miles)

›‹ 650 km (400 miles)

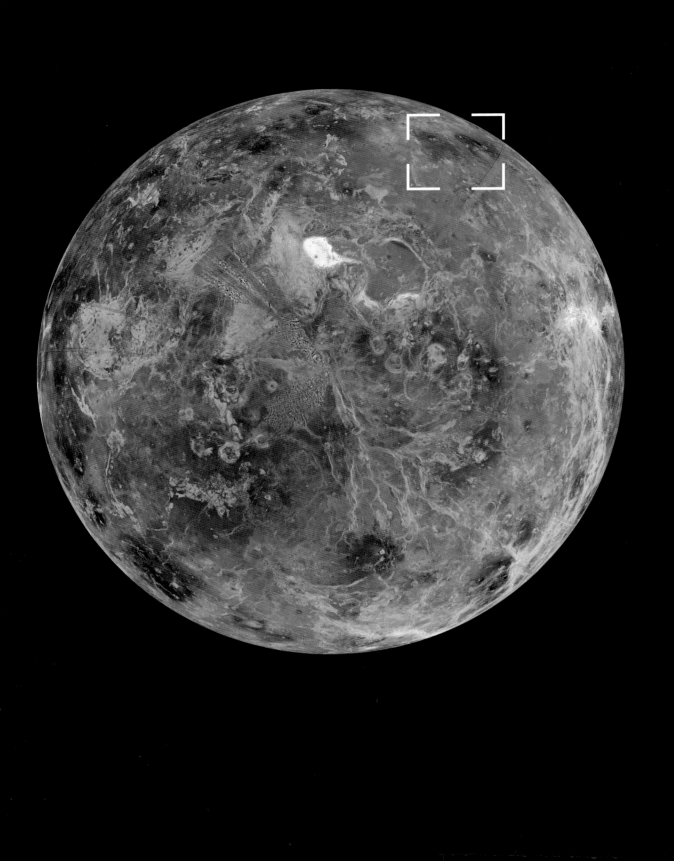

VENUS: A VOLCANIC WORLD

Despite being named after the ancient Roman goddess of beauty, Venus is in fact a hellish world. Veiled in a choking atmosphere dominated by deadly carbon dioxide, it has suffered from a runaway 'greenhouse effect' that traps heat close to the surface and ensures that temperatures rarely fall below 460°C (830°F). The atmosphere is so thick (exerting 100 times the pressure of Earth's at the planet's surface) that Venus' landscape is permanently hidden from space. A handful of robot probes have survived the descent through heat, pressure and corrosive acid rain to send back a few pictures of the surface, but our best view of Venus comes from radar maps produced by orbiting satellites.

SIF MONS

Venusian radar maps such as those from NASA's Magellan satellite reveal that Venus is a world dominated by volcanoes in many different forms. Most impressive are the towering shield volcanoes such as Sif Mons, shown here in a Magellan radar visualization that exaggerates vertical scale by a factor of ten in order to show the volcano's massive 300-km (186-mile) diameter more clearly. Surrounded by solidified lava flows, these volcanoes dominate raised plateaus known as Maxwell Montes and Alpha and Beta Regio.

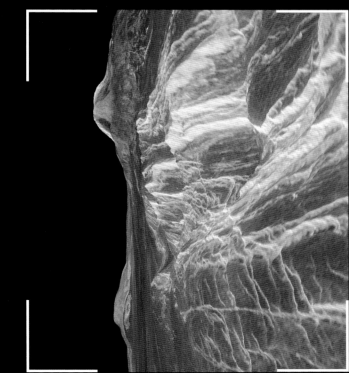

Craters on Venus are few and far between – partly because the planet's dense atmosphere shields it from space, but also because the surface is comparatively young compared to other terrestrial planets apart from Earth. According to most experts, Venus was largely resurfaced in a series of enormous volcanic eruptions that petered out around 500 million years ago. Only a few highland areas were left unaffected by this resurfacing, so any craters that are found elsewhere must have formed since.

■ LAVINIA PLANITIA

Venus's Lavinia Planitia lava plain is dominated by three enormous craters – Howe, Danilova and Aglaonice – ranging from 37 km (23 miles) to 62 km (39 miles) in diameter. Because incoming meteorites have to be a certain size in order to make it through the atmosphere without burning up, even the smallest Venusian craters tend to be at least 30 km (18 miles) across. This trio is thought to have formed when a single large meteorite broke up during the final stages of its descent. Because the atmosphere also limits the escape of 'ejecta' material as the crater forms, Venusian craters are often surrounded by obvious 'splashes' of ejecta.

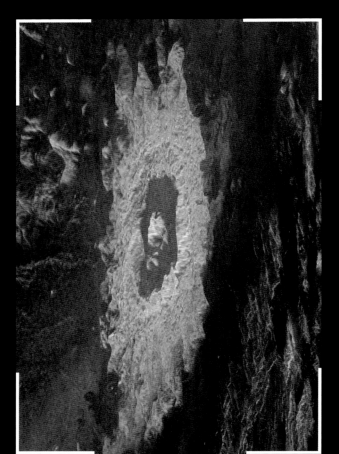

< > 77.3 million km (48 million miles)

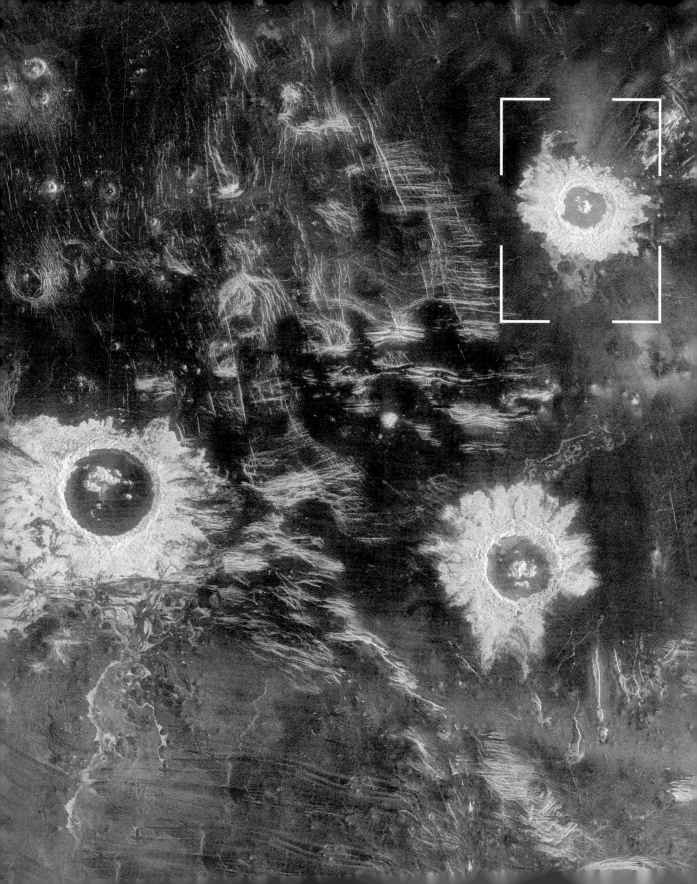

VENUS: EQUATORIAL RIFTS

Venus's diameter is almost 95 per cent of Earth's, allowing the planet to generate and trap large amounts of internal heat, so we might expect the two planets to show fairly similar geology, and perhaps even tectonic plates similar to those seen on our home planet. However, it seems that any attempt to form such plates on Venus rapidly ground to a halt, leaving fossilized traces in the form of deep rifts near the planet's equator. Astronomers think that the difference may be due to the lack of water on Venus – liquid water on Earth helps to 'lubricate' the slow movement of plates around our planet.

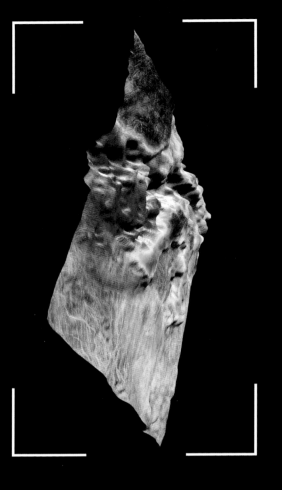

77.3 million km (48 million miles)

1,125 km (700 miles)

OVDA REGIO
This colour-coded 3D image of the equatorial highland region shows deep trenches running across an elevated plateau called Ovda Regio. Different colours indicate the way in which different parts of the terrain reflect back the Magellan satellite's radar signals – an indication of different mineral compositions. The central trench seems to consist of a series of parallel ridges and valleys, most likely formed as Venus's crust attempted to separate into individual tectonic plates early in the planet's history.

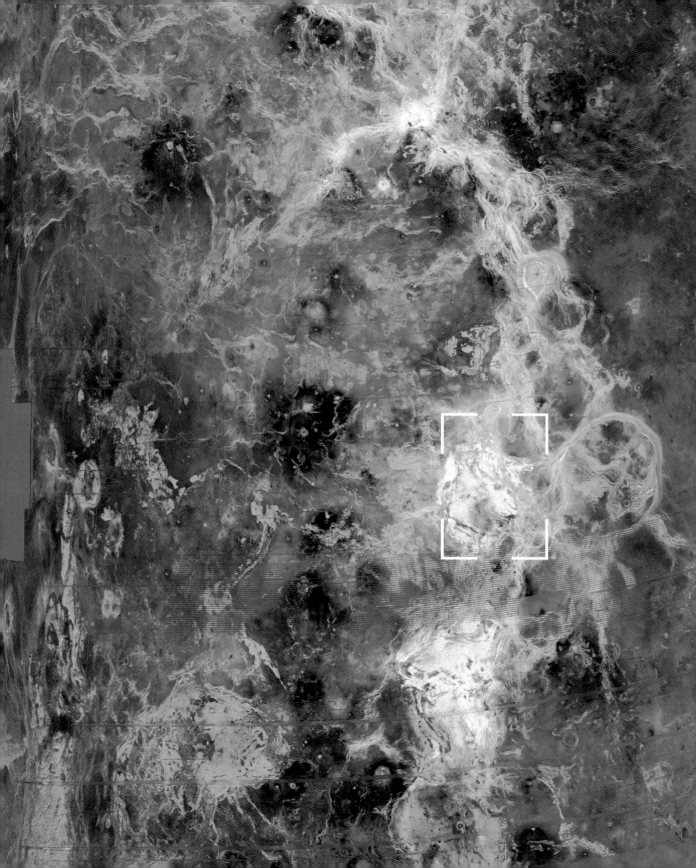

The Valles Marineris or Mariner Valley on Mars is one of the most spectacular features in the entire solar system – a scar across the Martian surface more than 4,000 km (2,500 miles) long and up to 10 km (6 miles) deep that dwarfs Earth's own Grand Canyon. Named after the Mariner 9 space probe that discovered it in 1972, the valley runs along the southern edge of a great bulge in the Martian surface known as the Tharsis Rise, formed from lavas laid down by long-lived volcanoes (see overleaf). The enormous weight of the bulge is thought to have caused huge stress that split the nearby surface apart.

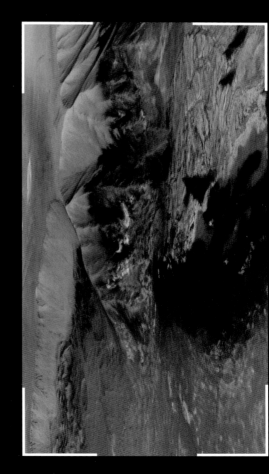

GRAND CANYON

This 3D view of the Melas Chasma region within the Valles Marineris was constructed using the European Mars Express satellite's stereo camera, which photographs features from slightly different angles. Melas Chasma is the broadest region of the canyon system, and studies of the minerals exposed on its cliff faces suggest it was formed by landslips in an area that was once covered by a large lake. The valley floor is the lowest-lying area of the Martian surface and therefore experiences the greatest air pressure, making it a favoured location for a possible future manned outpost.

18

MARS: OLYMPUS MONS

The towering peak of Olympus Mons is the biggest mountain in the solar system – an enormous 'shield volcano' made from lava that poured from multiple fissures along its flanks over hundreds of millions of years. Its dome is more than 620 km (385 miles) across, and rises to 27 km (17 miles) above the average Martian surface, or 19 km (12 miles) above the Tharsis bulge on which it rests. Some areas on the volcano's slope seem remarkably young – as little as 2 million years old – suggesting that Mars might still see occasional volcanic activity today.

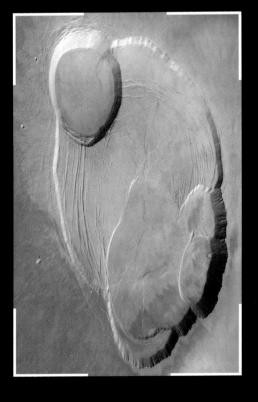

< > 55 million km (34.2 million miles)

∨ 85 km (53 miles)

■ COMPLEX CALDERA

The huge scale of Olympus Mons means that it is remarkably shallow – an astronaut walking on its flanks would barely notice the slope. At the peak, however, steep cliffs plunge more than 3 km (1.8 miles) into a complex central crater or 'caldera' – a series of interlocking rings created as the top of the volcano collapsed when the supporting pressure of hot rocks welling up through the Martian mantle was withdrawn at various times. From edge to edge, the caldera complex is more than 85 km (53 miles) across.

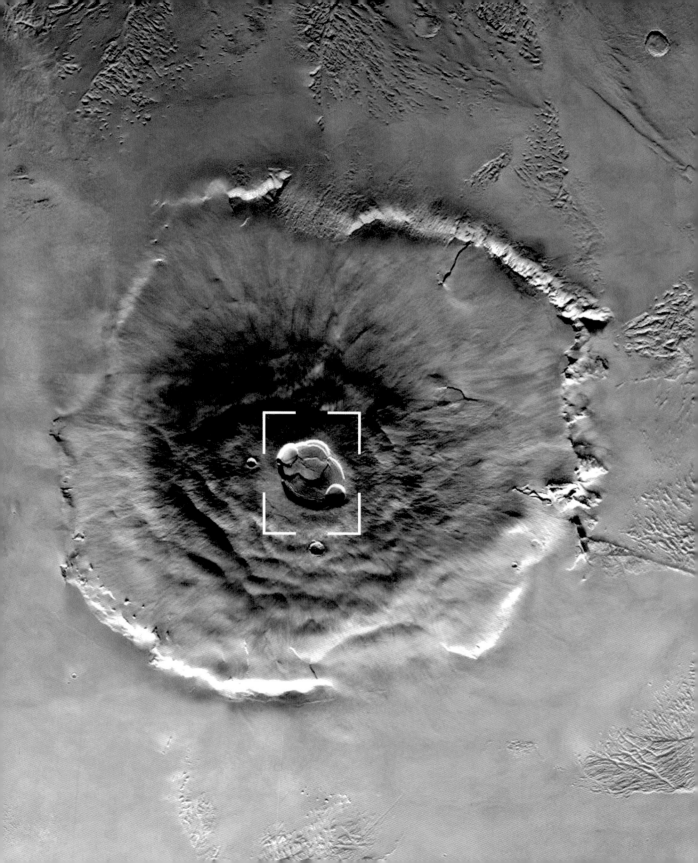

MARS: POLAR ICE

From a distance, the brilliant white polar caps of Mars form a stark contrast to the planet's general rusty red hues. Their existence was discovered in the 18th century, and they inspired a view of Mars as an Earth-like world with water flowing on its surface, and perhaps even artificial water channels built by intelligent Martians. The reality is rather different – the visible caps are largely composed of frozen carbon dioxide that falls out of the thin atmosphere or 'sublimes' as bright frosts that come and go with the Martian seasons. However, both poles have a more permanent underlying layer of frozen water ice.

ARCTIC LAYERS

Seen from orbit, the polar caps reveal beautiful swirling patterns around their edges. These are caused by the way in which carbon dioxide frosts repeatedly sublime and evaporate from year to year. Each year's frost contains impurities in the form of fine Martian dust, and these are left behind when the frost itself evaporates. Deposits concentrate in specific areas while others are eroded away as a result of prevailing winds caused by the planet's daily rotation. The result – complex plateaus rising up to 3 km (1.8 miles) above the Martian surface.

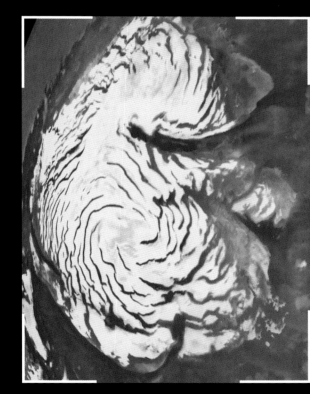

MARS: VICTORIA CRATER

At 730 m (2,400 ft) wide, Victoria crater is a relatively insignificant feature in the landscape of the Meridiani Planum region close to the Martian equator. However, it is one of the most intensively studied regions of the red planet's surface thanks to its location close to the landing site of the NASA's Opportunity Mars Rover. Photographs from the Mars Reconnaissance Orbiter satellite have helped to put Victoria in context – they reveal a crater edged with numerous landslips and filled with fine Martian dust that piles up in dunes near the centre of its floor.

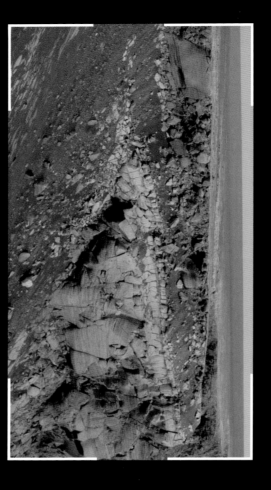

55 million km (34.2 million miles)

30 m (100 ft)

■ CAPE ST. VINCENT
Martian craters are important targets for rovers such as Opportunity because they act as ready-made excavations into the planet's crust. Coupled with satellite observations, they have confirmed the existence of sedimentary rocks (formed from the gradual deposition of sandy deposits from standing lakes or seas) across large areas of the Martian surface. Opportunity and its twin rover Spirit have discovered minerals that could only have formed in water, increasing the evidence that the dry, cold planet Mars we see today was once a warm, wet world.

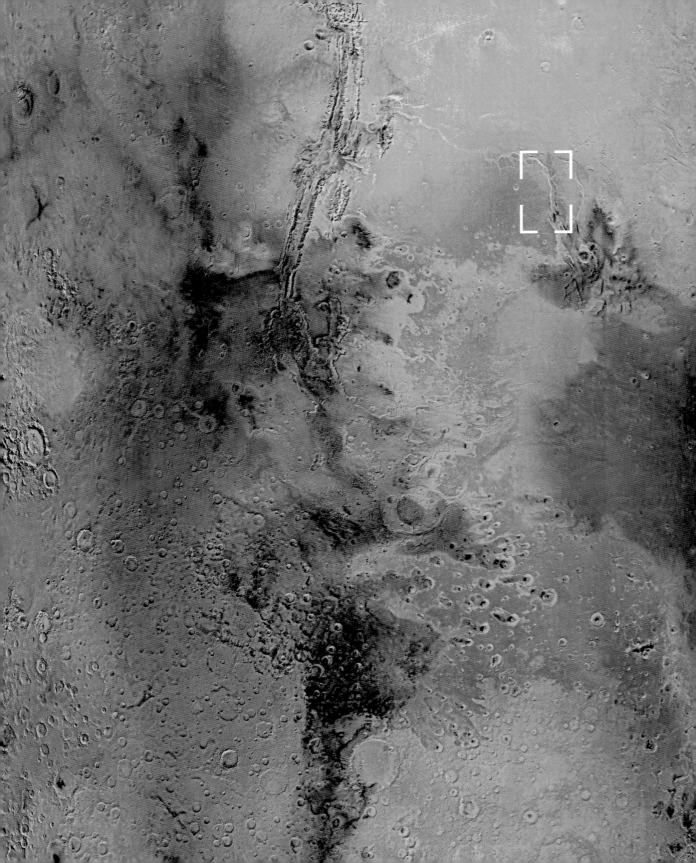

21

MARS: FLOODPLAINS AND RIVERS

The landscape of Mars can be broadly divided into two halves – heavily cratered southern highlands, and smoother northern lowlands (with the Tharsis bulge rising out of them). Some scientists believe that a catastrophe early in Martian history resurfaced the northern hemisphere, destroying the evidence of its early craters. West of Tharsis and directly on the north/south boundary lies a circular plain known as Chryse, which still bears teardrop-shaped scars from catastrophic floods, as well as numerous winding channels resembling river valleys. It seems that, billions of years ago, Chryse formed one of the main 'outflows' for water accumulating in the southern highlands and flowing out of the Valles Marineris.

KASEI VALLIS

Situated on the edge of the Chryse floodplain, Kasei Vallis is an enormous outflow channel from the Martian highlands, up to 300 km (186 miles) wide, and with edges that show numerous gulley-like structures. Studies of similar gulleys from elsewhere on Mars have shown that they can be remarkably young – suggesting that perhaps Mars is not as cold and dry as it seems, and that water may still survive in liquid form in underground reservoirs, occasionally bursting to the surface before it boils away into the atmosphere.

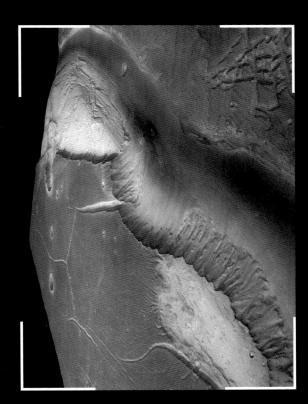

55 million km (34.2 million miles)

300 km (186 miles)

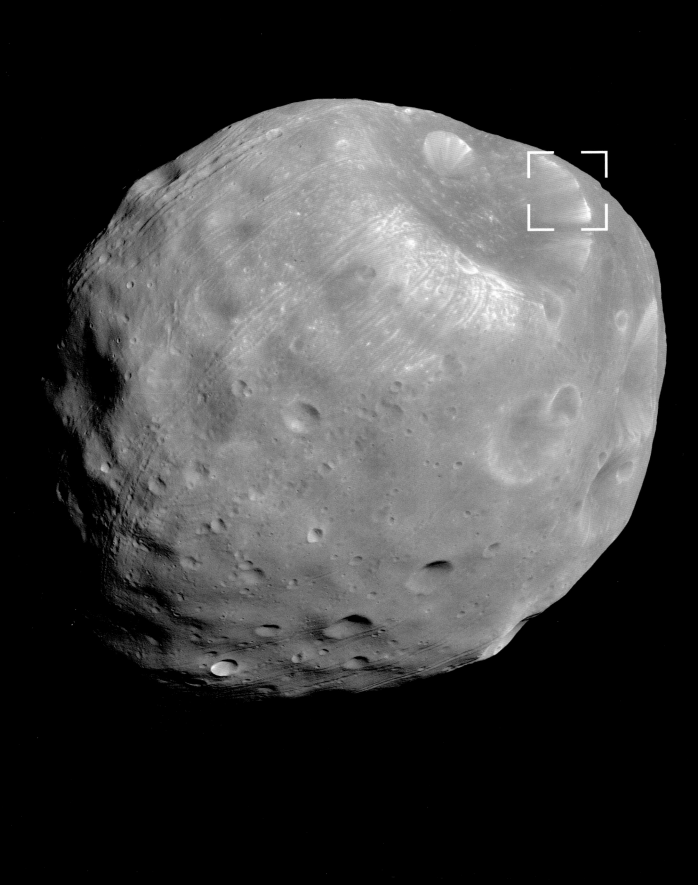

Mars is circled by two small moons, named Phobos and Deimos after the sons of Mars in classical mythology. Each is much smal[l] than Earth's own Moon – Phobos, the innermost and larger of the two, is just 22 km (14 miles) across, and both orbit much close[r] to their parent planet than Earth's Moon, circling Mars in just 7.7 hours and 30.3 hours respectively. Despite its asteroid-like features, it now seems that Phobos is not a captured asteroid, but is instead a chunk of debris ejected from Mars itself during a huge ancient impact.

■ STICKNEY CRATER

Phobos's most distinctive feature is a huge crater called Stickney, some 9 km (5.6 miles) across. This close-up view of the crater wall from the Mars Reconnaissance Orbiter space probe shows a distinctive 'splash' pattern up its walls. Despite appearances, the 'splash' does not seem to have been made by material thrown out from the crater, but instead by downward landslips towards the crater floor.

Not all asteroids orbit strictly between Mars and Jupiter – many come within the orbit of Mars for at least part of their orbit, and these 'Near Earth Asteroid' are often the easiest to study. 433 Eros, whose orbit ranges from just outside Earth's orbit to just within that of Mars, is probably the most intensively studied asteroid of all. It was circled by the Near-Earth Asteroid Rendezvous space probe for almost a year. Appropriately, considering Eros's name, NEAR arrived in orbit on Valentine's Day 2000.

■ POWDERY SURFACE

As NEAR continued its year-long orbit of Eros, it slowly spiralled inwards and sent back increasingly detailed images of the surface. Despite its present orbit, Eros must have originated in the main asteroid belt, where it would have been subjected to regular collisions, so astronomers were surprised to find that much of its surface was relatively smooth. Eventually they concluded that a large collision roughly a billion years ago had produced shockwaves that pulverized much of the surface and caused many of the smaller, older craters to collapse.

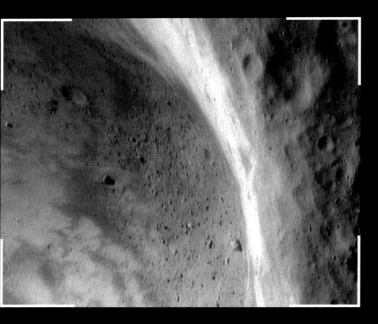

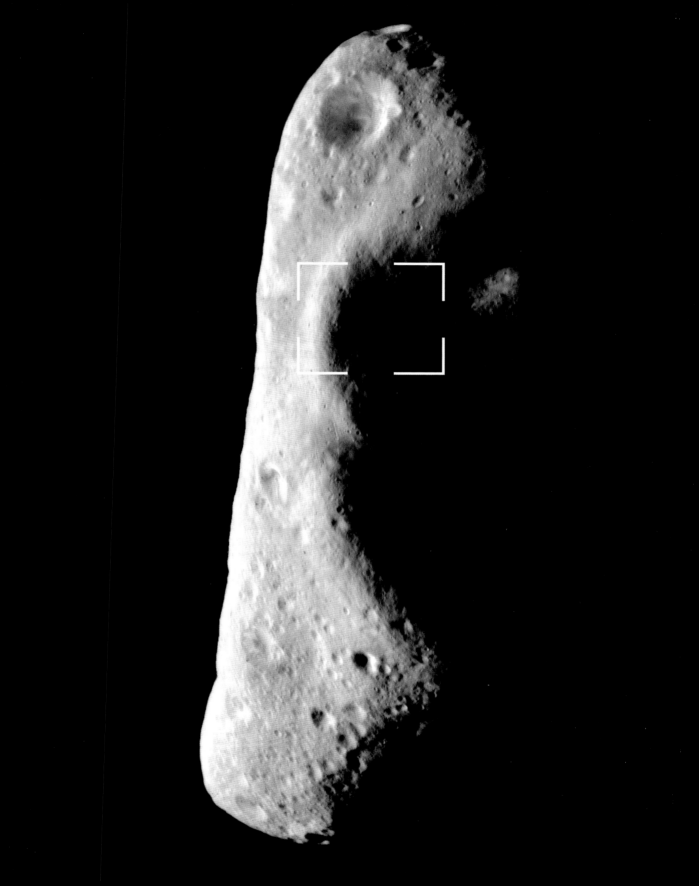

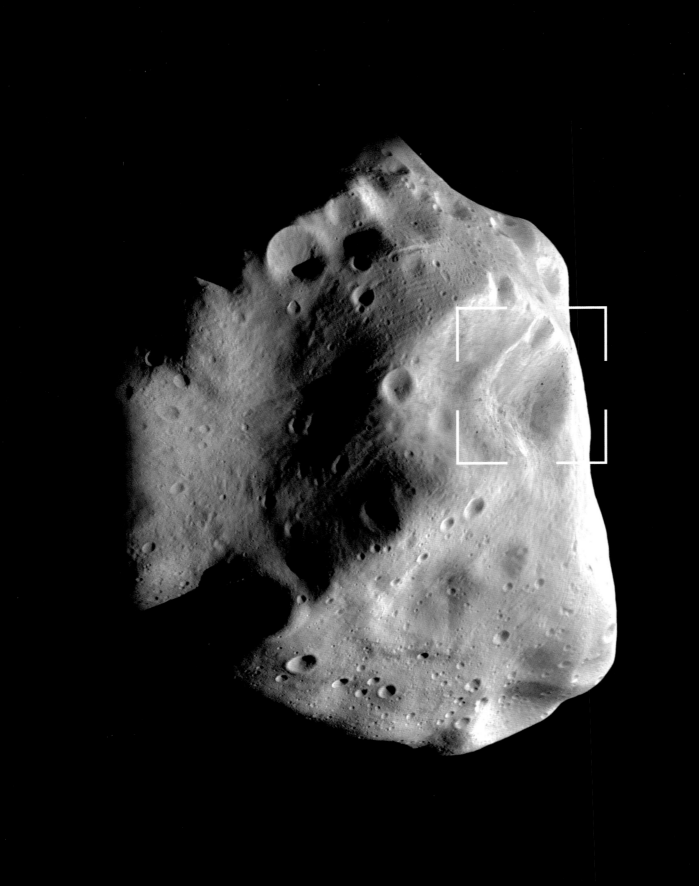

ASTEROID 21 LUTETIA

The largest asteroid so far visited by a space probe is technically known as 21 Lutetia (since it was the 21st asteroid to be discovered and catalogued, in 1852). This large misshapen rock, some 132x101x76 km (82x63x47 miles) across, has an unusual surface composition, and so was targeted for investigation by the European Space Agency's comet probe Rosetta, on its way to its ultimate destination in July 2010. Most asteroids are either C-type, believed to be unaltered material from the birth of the solar system, S-type (showing chemically altered minerals) or M-type, rich in metals and believed to be the broken-up cores of earlier bodies. Lutetia seems to be halfway between C-type and M-type.

■ MAJOR CRATER

Despite its size, Lutetia proved to have features typical of many asteroids so far visited by space probes. The surface is heavily cratered and uneven, with signs of at least one large 'dent', probably caused by a collision with a substantial neighbour early in its history. In other places, the landscape looks relatively smooth, probably due to a thin coating of impact-created dust filling in small-scale craters. Among the puzzling features are the parallel scratches found on some parts of Lutetia – similar marks have been seen on Mars's small satellites Phobos and Deimos.

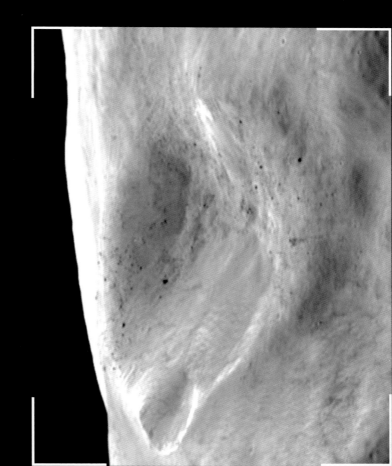

Jupiter is the giant of the solar system, and the contrasting colours of its atmosphere give it an unmistakable striped appearance. Early telescopic astronomers were puzzled by these patterns until they realized that the planet's visible surface is just the top of a deep atmosphere – one that we now know comprises the vast majority of the planet, with perhaps just a small solid core at the centre. The vast majority of the planet is pure hydrogen, which condenses into liquid form under pressure a few thousand kilometres below the surface, and deeper still breaks up into a 'metallic' form that generates a powerful electromagnetic field around Jupiter.

■ BANDED WORLD

The dominant features of the Jovian atmosphere are the cloud bands that are wrapped parallel to Jupiter's equator by the planet's ten-hour rotation. Bright bands are called zones, and dark stripes are known as belts. Although this terminology suggests that the belts lie on top of the zones, the reality is precisely the opposite. Zones mark regions of low pressure where clouds 'pile up' and creamy ammonia crystals condense high in the atmosphere. Belts, meanwhile, are high-pressure clearings that provide a view into deeper layers of the atmosphere containing complex chemicals.

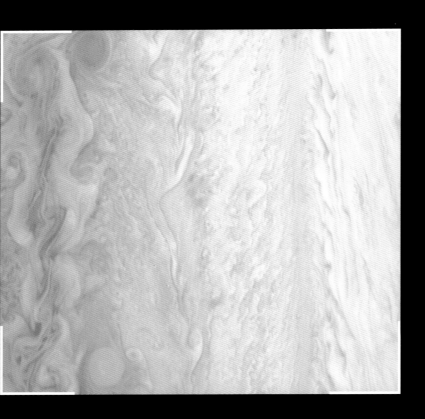

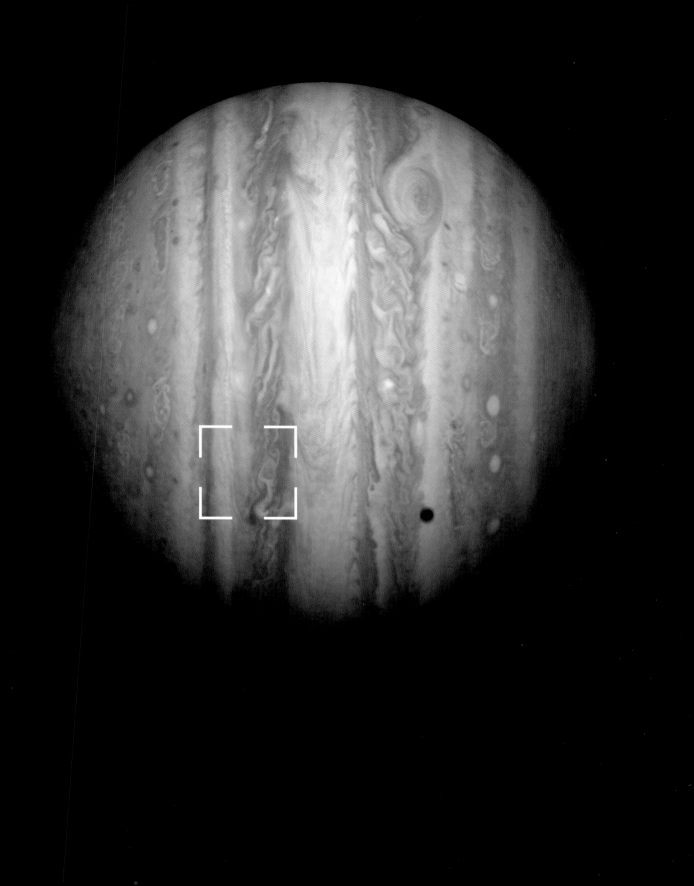

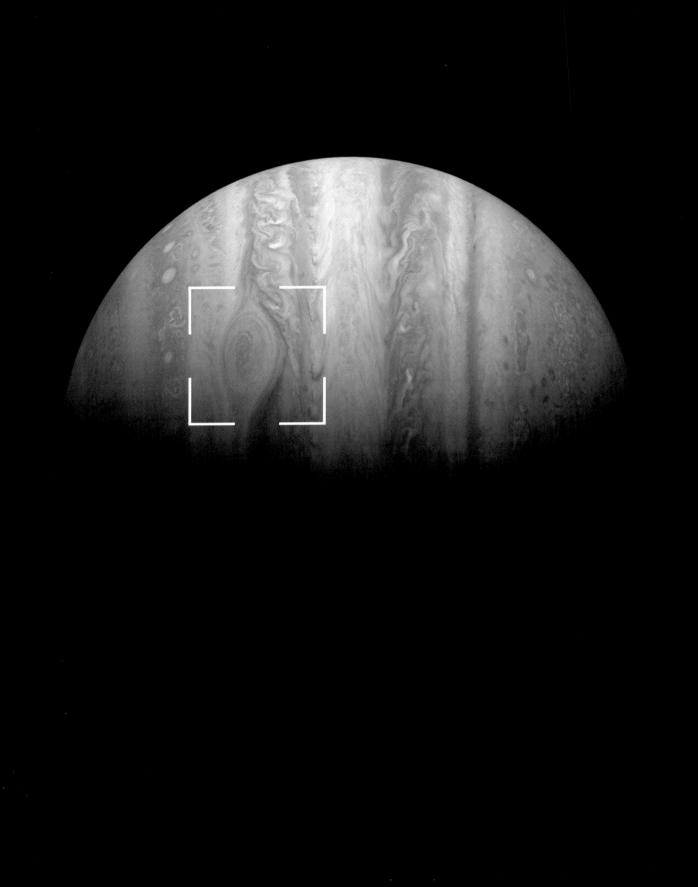

Jupiter's most famous feature is a vast oval storm large enough to swallow the Earth, swirling anticlockwise at a latitude of about 22° South of the Jovian equator. Known as the Great Red Spot, or GRS, this storm has been observed for at least 180 years, and perhaps for almost 350 – astronomers once thought it was an enormous island floating in a global ocean. Its colour can vary hugely over time and on occasion it can completely disappear (leaving a 'hollow' in its place), but it always seems to regenerate itself, seemingly by swallowing up other, smaller storms.

■ SUPERSTORM

The GRS is thought to be an enormous high-pressure area that dredges up material from deep within the planet, causing brightly coloured clouds to condense at high altitudes. It towers up to 8 kilometres (5 miles) above the surrounding clouds, and rotates once every six Earth days. Despite centuries of study, astronomers still aren't sure what causes its colour – it may be due to phosphorus, sulphur compounds, or complex carbon-based 'organic' molecules.

In the past decade, astronomers have noticed the formation of two separate smaller red spots in Jupiter's atmosphere. The first and largest of these, technically termed 'Oval BA' but often known as Red Spot Jr, formed from three white storm ovals that had been observed for several decades circling the planet south of the GRS. These three merged together in 1998 and 2000 to form a single large white storm, which began to turn red in 2005. Since then, Oval BA has strengthened further, growing to the size of the Earth (about half the size of the GRS) and increasing its rotation speed.

■ BABY RED SPOT

In early 2008, a third red spot appeared in Jupiter's southern cloud bands. Known as the South Tropical Little Red Spot (LRS), this storm had previously been observed as a bright white oval, circling the planet at more or less the same latitude as the GRS. In May 2008, it began to intensify and develop a strong red colour. However, the LRS was doomed to a short life – barely a month later, during a close encounter with the GRS and as Oval BA passed by, it was shredded to pieces, with its constituent parts circling the larger spot for a short time before being absorbed completely.

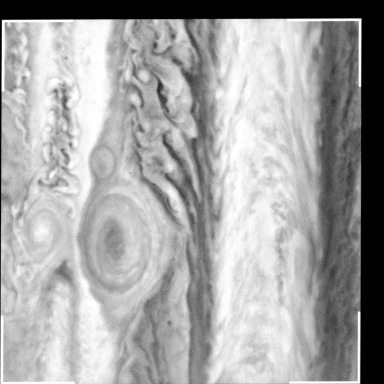

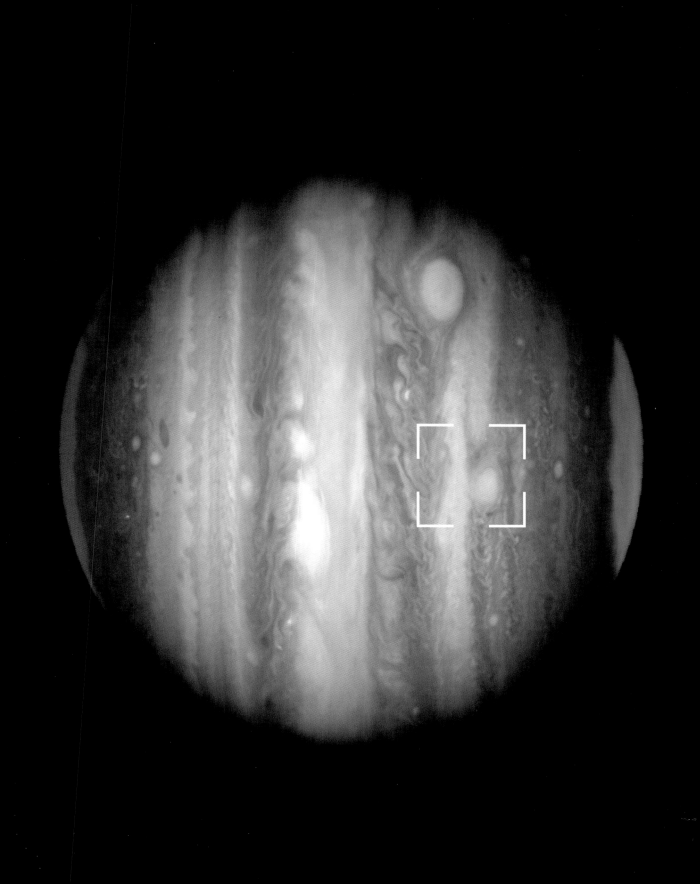

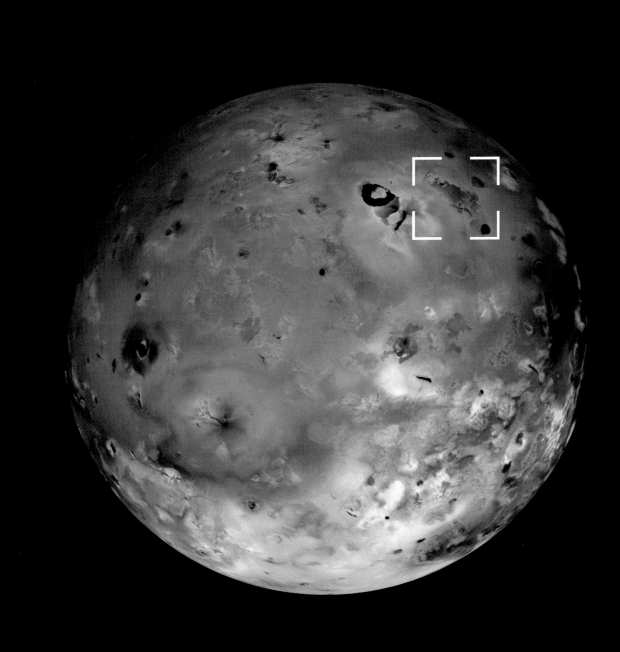

IO: LOKI VOLCANO

The innermost of Jupiter's four large 'Galilean' moons (discovered by Italian astronomer Galileo Galilei in 1610), Io is the most volcanically active world in our solar system. Gravitational forces from Jupiter pull it in different directions throughout each 42-hour orbit, heating the interior and powering widespread eruptions. Io's volcanism is boosted considerably by the fact that much of the surface rock is made of sulphur compounds that melt at low temperatures. Different forms of sulphur known as allotropes can take on a wide range of colours, predominantly red and orange, yellow, green and white, giving Io a surprisingly colourful surface.

SULPHUR PLUMES

Volcanoes on Io take a variety of forms. In places, lava lakes and rivers melt away the surrounding rock, while in others, such as Loki shown here, molten sulphur trapped beneath the surface bursts through and erupts into the sky in towering geysers. Droplets of sulphur condensing within these fountains may fall back to help coat the surface, or escape Io's weak gravity entirely, entering orbit around Jupiter itself. Here, funnelled by the giant planet's powerful magnetic field, they form a doughnut-shaped ring 'torus' around Io's orbit.

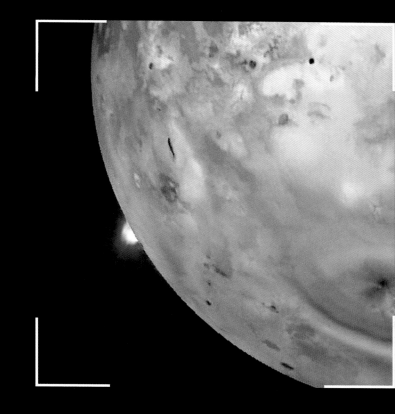

[29] IO: BOOSAULE MONTES

Io has some of the highest mountains in the solar system, reaching altitudes of up to 17,500 m (57,400 ft). Unusually for extraterrestrial mountains, though, they are not volcanic in origin – instead they are formed in a similar way to many of Earth's mountain chains, through 'tectonic' shifts in the moon's crust. Unlike Earth, Io's crust is not split into distinct plates, but currents in the molten rock within the moon are still strong enough to pull the crust around and cause it to pile up in places like a rumpled rug.

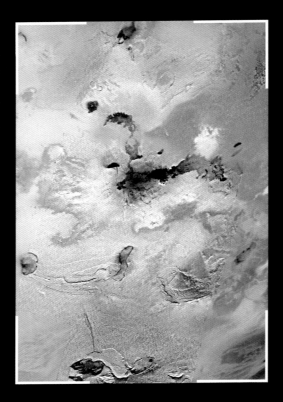

816 million km (507 million miles)

550 km (340 miles)

■ WRINKLED PEAKS

The Boösaule Montes mountains, Io's tallest, form a huge ridge some 550 km (340 miles) long. Discovered by the Galileo probe that orbited Jupiter between 1995 and 2004, their enormous height shows they cannot be volcanic – the sulphurous lavas produced by Io's volcanoes could not build up into such a heavy structure without giving way. Conversely, the mountains also show that Io's crust is stronger than suspected – the underlying rocks must contain considerable amounts of Earth-like silicate minerals in order to support them.

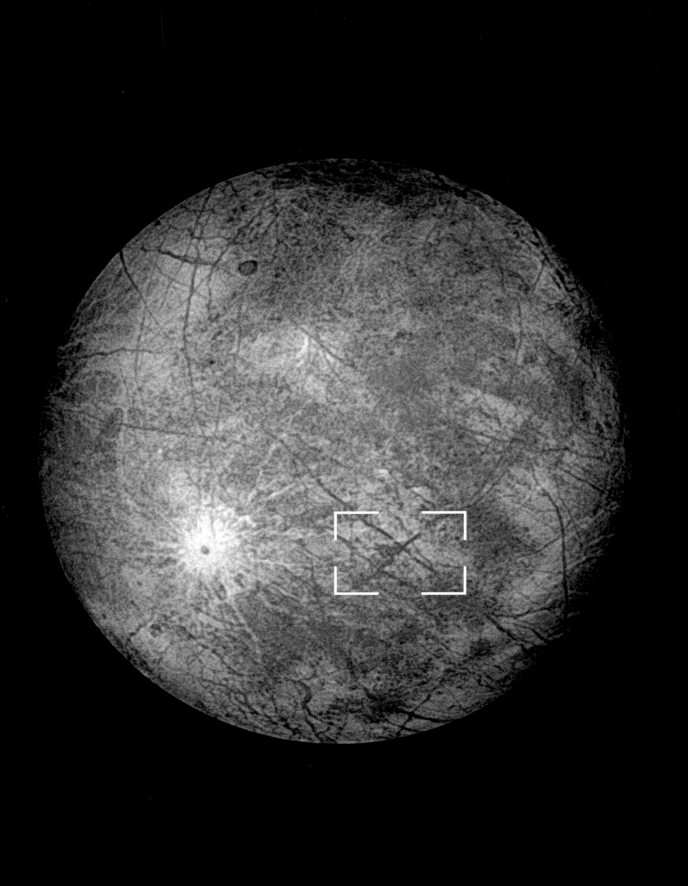

EUROPA: ICY CRUST

At first glance, Jupiter's second major moon, Europa, forms a stark contrast to the turbulent Io. It appears to be a placid, frozen world, covered in brilliant pinkish-white ice. However, this kilometres-thick icy crust hides an amazing truth – deep down, Europa is a waterworld covered with a global ocean of liquid water perhaps 100 km (62 miles) deep. The water is kept warm by volcanic activity on the seafloor, similar to that seen around deep-sea trenches on Earth, and many astronomers hope that Europa might be an ideal place to look for alien life in our solar system.

ANCIENT FRACTURES

Enhanced-colour images of Europa's surface reveal that it is criss-crossed by dark lines and complex patterns – the moon's entire crust bears a striking resemblance to the pack ice found near Earth's own poles. Although Europa's crust is far thicker than Earth's sea-ice, it still seems to be constantly reshaping itself, pulling apart in some places, thickening and remelting in others. The dark marks are thought to be caused by upwellings of warmer ice, coloured by nutrient-rich water from the ocean beneath.

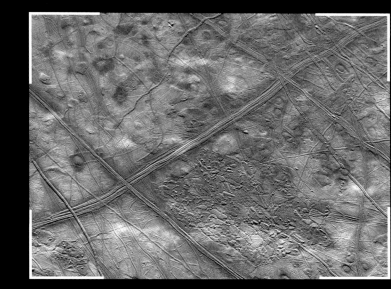

EUROPA: MOBILE SURFACE

Europa's surface is pool-ball smooth – if the moon were expanded to the size of the Earth, it would still have no hills taller than about 200 m (660 ft). It seems that the icy crust is not truly solid – instead it behaves rather like a glacier or ice sheet on Earth, slumping and flattening out to erase elevated features on a relatively short timescale (perhaps thousands of years), leaving only their ghostly remnants behind. Combined with the general resurfacing suggested by the crust's 'ice raft' appearance, this makes Europa's visible surface one of the youngest and freshest in the solar system.

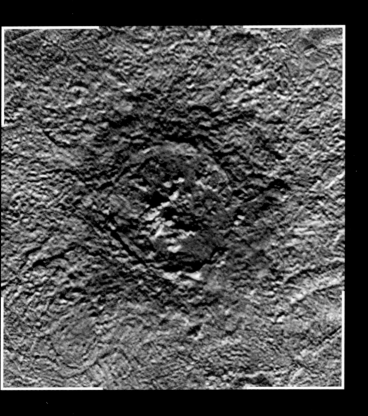

◼ PWYLL CRATER

Only a handful of impact craters are known to exist on Europa, and Pwyll is among the youngest and best defined. The central crater is about 26 km (16 miles) in diameter, while the spray of white ejecta surrounding it extends over hundreds of kilometres and clearly overlies everything else in the vicinity. The ejecta is likely to be pure water ice, vaporized and then condensed back into snow by the energy of the impact, while the dark crater floor suggests that Europa's surface has already begun to 'heal' the scar through upwelling of impure ice from below.

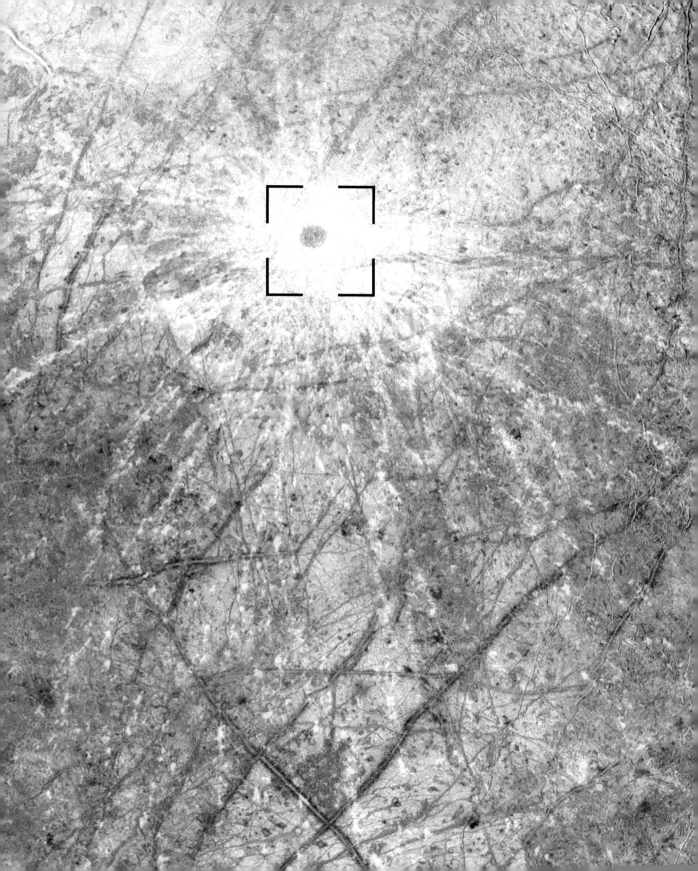

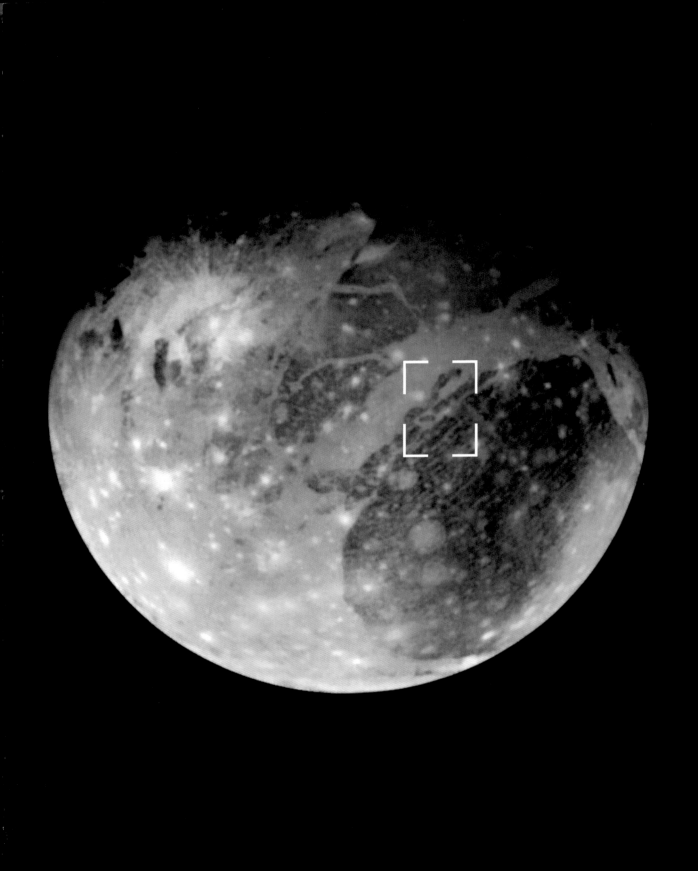

The third of Jupiter's Galilean moons, Ganymede, is also the largest moon in the solar system – a world that is larger than Mercury. At first glance, it seems disappointingly inactive compared to its inner neighbours, with a frozen surface made from a mix of ice and rock, and no substantial atmosphere. However, Ganymede's orbit has evolved over the 5-billion-year lifetime of the Jupiter system, and it, too, was once in a position to receive significant amounts of 'tidal heating' from Jupiter. Residual heat from this time may even help to maintain a thin saltwater ocean just beneath the visible crust.

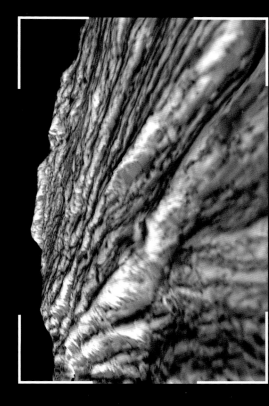

■ URUK SULCUS

Ganymede's surface is a patchwork of light and dark regions, with a passing resemblance to Earth's continents. The number of craters that have accumulated in different parts suggests that the dark regions are significantly older than the light ones. The boundaries between these different terrains are often marked by long strips of parallel ridges known as sulci; the best photographed of which is Uruk Sulcus. They seem to mark places where new crust has been formed as the older surface stretched apart and warm fresh ice forced its way up through the cracks.

Jupiter's gravity exerts a powerful influence on asteroids, comets and other fragments of interplanetary debris in its vicinity, and makes its satellites sitting ducks for infalling objects. While the activity on Io and Europa rapidly wipes away the traces of impact, Ganymede's older surface maintains a much more detailed record. The most likely explanation for the difference between darker, heavily cratered regions and brighter, less scarred areas is that Ganymede's entire ancient surface broke into fragments at some point in its history, with some regions sinking back into the moon's interior and being replaced by a fresh upwelling mix of rock and ice.

■ ENKI CATENA

The chain of impacts known as Enki Catena is probably Ganymede's most famous crater group. A catena is a chain of craters, and Enki Catena displays some 13 neatly aligned craters, each about 16 km (10 miles) across, stretched out across 160 km (100 miles) of the moon's surface. The impacts were probably formed when Ganymede's path crossed that of a fragmented comet on a collision course with Jupiter – they are strangely reminiscent of the famous comet Shoemaker-Levy 9, which fragmented in just this way after a close encounter with the giant planet and eventually slammed into Jupiter's atmosphere, unleashing spectacular explosions, in 1994.

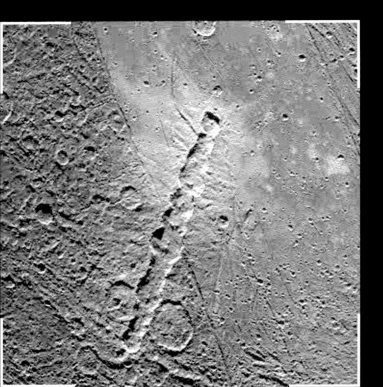

CALLISTO: CRATERED MOON

While Jupiter's inner Galilean moons have all been shaped by the influence of 'tidal heating' from the giant planet itself, Callisto – the outermost and second largest – has escaped these effects. As a result, its surface retains a pristine record of more than 4 billion years of impacts from objects drawn in by Jupiter's gravity – experts believe it is the most cratered world in the solar system. Callisto's outer crust is naturally quite dark, but each new impact sprays out fresh ice from just below the surface, giving the moon the overall impression of a cosmic glitterball.

RINGED BASIN

Valhalla is the largest impact structure on Callisto – a huge crater bounded by a dozen or more concentric rings. According to some measures its overall diameter extends to 3,800 km (2,360 miles), making it easily the largest crater in the solar system. Compared to many smaller craters, the centre of Valhalla is made up of comparatively bright material that is believed to have welled up from within the moon to heal the enormous scar left on its surface. Astronomers call such regions on icy moons 'palimpsests' (originally a term for a reused piece of ancient parchment overwritten with fresh content).

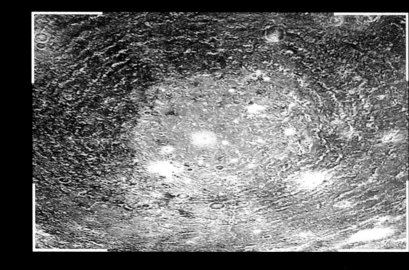

SATURN: ATMOSPHERE

In contrast to its inner neighbour Jupiter, Saturn itself appears to be a placid, well-ordered world – its cream and white cloud bands are very different from the colourful turbulence of Jupiter. However, the planets are more similar than first appearances suggest. Cooler conditions further from the Sun combine with Saturn's weaker gravity, allowing clouds to condense at greater heights and wrap the entire planet in a creamy haze of ammonia. Saturn, like Jupiter, is a gas giant with a huge atmosphere wrapped around a small solid core. Its rapid rotation and low density cause it to bulge noticeably as fast-moving material around the equator attempts to escape completely.

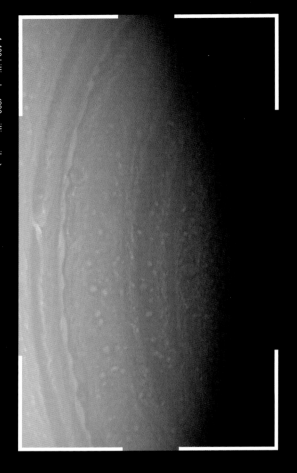

1.433 billion km (890 million miles)
71,000 km (44,000 miles)

■ WINTER BLUES

Since 2005, Saturn has been studied in detail by NASA's Cassini orbiter. With a tilted axis similar to that of Earth, the ringed planet goes through a similar cycle of seasons – though stretched over a Saturnian year equivalent to 29.5 Earth years. Currently, Saturn's northern hemisphere is warming up as it enters spring, and this beautiful natural-colour image shows a region close to the planet's north pole. Several distinct, well-defined cloud belts and zones can be clearly seen, along with a number of small bright storms. The blue tint to the poorly lit polar region is a result of sunlight scattering in the atmosphere – the same effect that makes Earth's own sky blue.

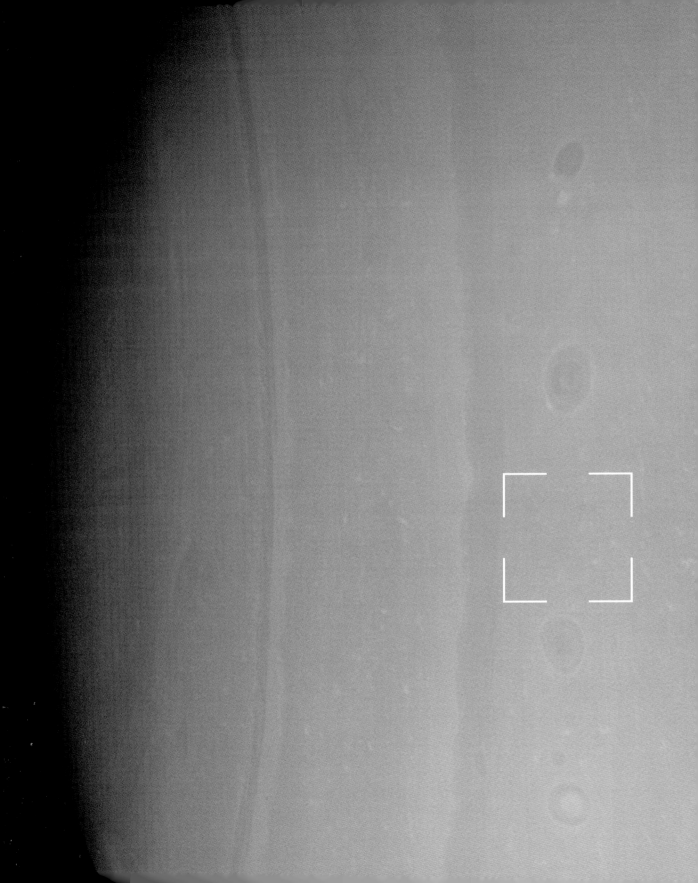

SATURN: STORMY WEATHER

As Cassini entered orbit around Saturn in 2004, mission scientists detected mysterious bursts of radio 'static' that came and went in a period matching the planet's 10.6-hour rotation. In 2005, a false-colour image (shown below), which peers into the depths of Saturn's atmosphere to study the planets emission of near-infrared 'heat radiation', revealed the culprit – a swirling feature buried to the south of Saturn's equator, and dubbed the 'Dragon Storm'. Colour within this image indicates depth within Saturn's atmosphere – grey features are near the surface, while red ones are further down. The storm generated lightning bolts 10,000 times more powerful than any on Earth, creating the static bursts.

▦ THE DRAGON STORM

The Dragon Storm lay beneath a band of Saturn's atmosphere known as 'Storm Alley', where visible bright storms frequently appear at high levels in the atmosphere (dark spots in the image). It faded away after several months, but has been succeeded by newer and longer-lived storms. The Dragon and its followers are thought to be a deep 'convective' thunderstorm, fuelled by an updraft of warm, moist material from within the atmosphere, which condenses to create torrential rainfall and lightning. However, it seems the energy of these buried thunderstorms can sometimes be transferred to the upper atmosphere, spawning bright white plumes that wrap themselves into spinning storms and gradually drift away from their originator.

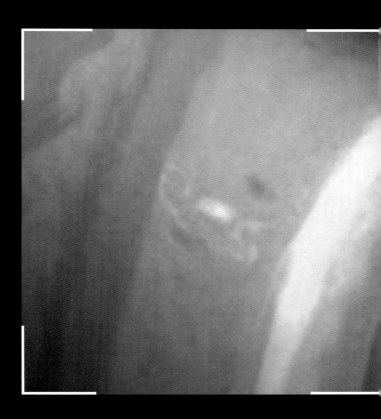

SATURN'S RINGS: FINE STRUCTURE

Saturn's most famous feature, of course, is its spectacular ring system. While all the giant planets have rings, Saturn's are by far the brightest and best developed. Different sections of the rings are easily distinguished from one another – the bright A and B rings are separated by an apparently empty gap called the Cassini Division, while the C (or 'Crepe') ring, inside the B ring, has a semi-transparent appearance. Finer detail includes the narrow Encke Division within the A ring; the thread-like F ring along its outer edge and the tenuous D ring stretching from the inner edge of the C ring down to the top of Saturn's atmosphere.

B-RING DETAILS

Viewed up close by space probes such as Cassini, the structure of the rings is revealed to be even more complex. Each major ring is composed of countless concentric ringlets like the grooves on a vinyl record – and even the apparently empty divisions turn out to contain a few. The distribution of ringlets seems to be governed in part by the influence of Saturn's countless moons – the individual particles that make up the ringlets tend to avoid 'resonant' orbits where they would suffer repeated close 'tugs' from a moon's gravity (for instance, orbits where they circle Saturn exactly twice as fast as a particular moon).

38

SATURN'S RINGS: PARTICLES

As early as 1859, Scots physicist James Clerk Maxwell realized that the rings of Saturn could not possibly be solid structures, and must instead be composed of countless billions of tiny particles, each following an individual orbit around the planet. However, it's only in the past few years that we've been able to glimpse such particles directly. Astronomers are still divided on the origin of the particles – the break-up of infalling comets, the break-up of an ancient moon and the gradual 'sandblasting' of fragments from the present-day moons may all play a part. Once in orbit, constant 'jostling' from other particles plays a key role in keeping each fragment in a concentric orbit and a narrow plane.

■ A PROPELLER IN THE RINGS

While the vast majority of ring particles are a few metres across or less (their size is one key factor in determining the appearance of the different rings), there are also countless small 'moonlets', up to a few hundred metres across, embedded within the rings. The influence of these moonlets gives rise to short-lived larger clusters of material, often in the shape of an aircraft propeller. Cassini snapped this striking image of one such 'propeller' casting its shadow across the A ring in September 2009. The propeller casting the shadow is roughly 130 km (80 miles) long, and the moonlet at its heart is probably about 400 m (1,300 ft) across.

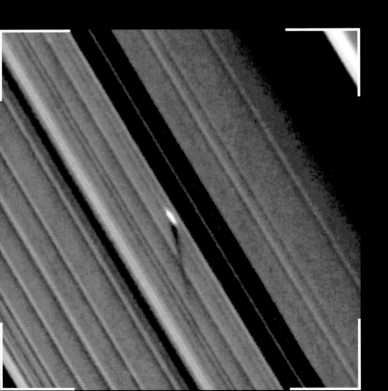

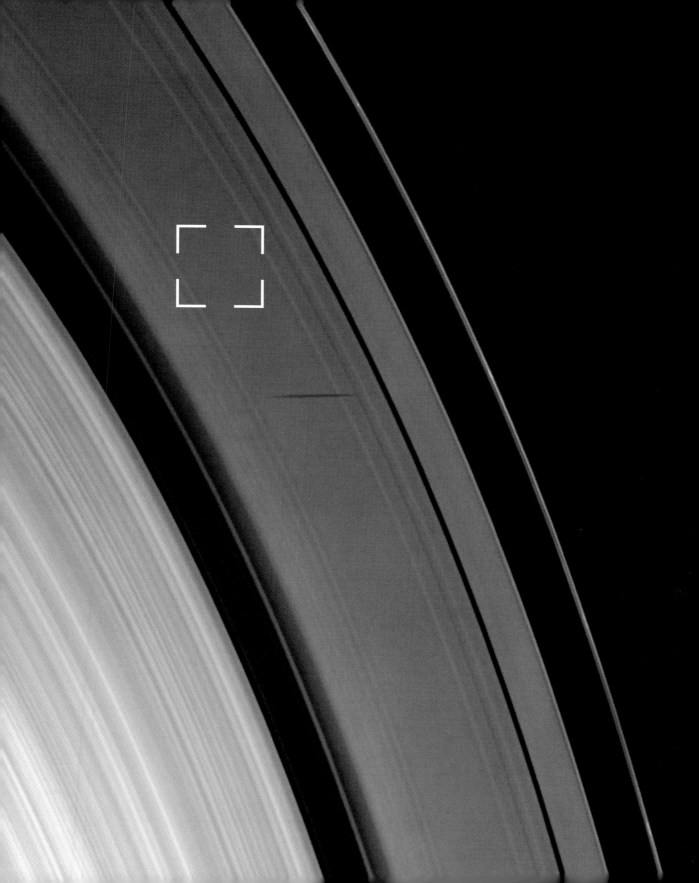

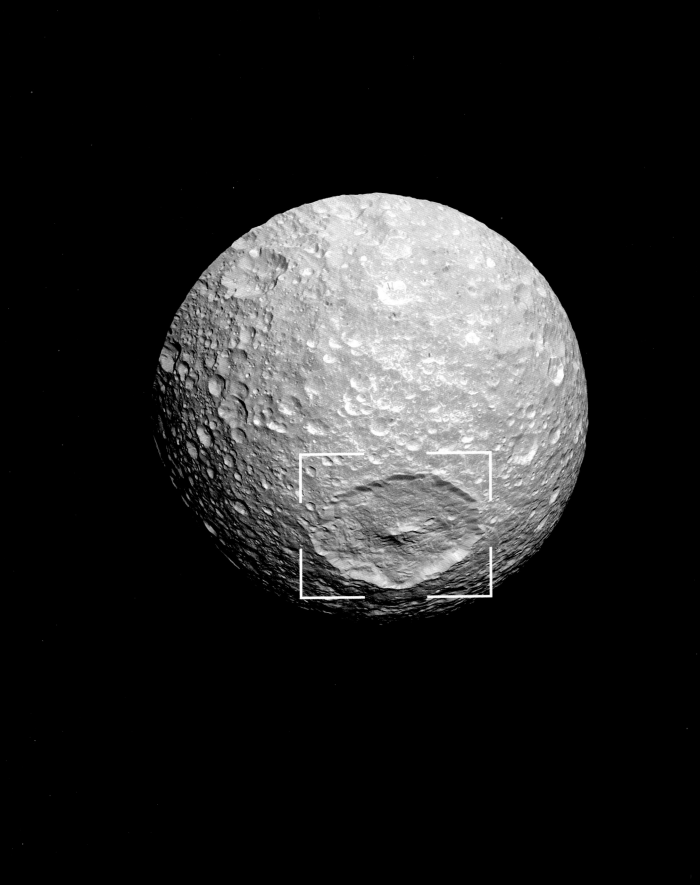

Saturn has a huge family of moons that display a great variety of surface features. The innermost satellites are small 'shepherd moons' that orbit in and around the rings themselves. Mimas is the innermost of the somewhat larger 'mid-sized' moons. With a diameter of just 400 km (250 miles), it is also the smallest world in the solar system with sufficient gravity to pull itself into a spherical shape. This is largely due to its composition – measurements of its density show it close to that of pure ice, which is much easier to pull into shape than solid rock.

HERSCHEL CRATER

Mimas is covered in craters, but the largest and most impressive by far is the Herschel crater, named in honour of the great astronomer William Herschel, who discovered Mimas in 1789. With a diameter of around 130 km (80 miles), Herschel dominates an entire hemisphere of Mimas's surface, and gives it an irresistible resemblance to the Death Star of the *Star Wars* movies. The impact that formed this huge crater must have been almost large enough to shatter the entire moon into fragments.

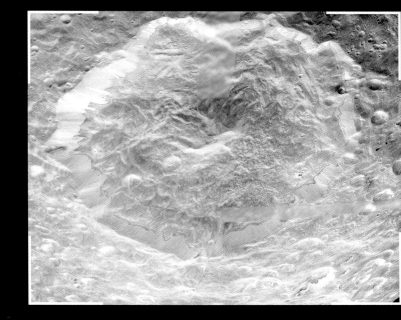

40

ENCELADUS: FROSTED WORLD

The second large moon of Saturn, Enceladus, is barely 100 km (62 miles) wider than Mimas, but is a remarkably different world, and one of the most fascinating moons in the solar system. Enceladus has a brilliant white surface – the brightest landscape in the solar system – and a distinct lack of craters. In fact, the entire moon is covered in relatively fresh snow, released in geyser-like eruptions from liquid water reservoirs just below the surface. This remarkable activity is driven by tidal heating caused by a gravitational 'tug of war' between Saturn and the outer moon Dione.

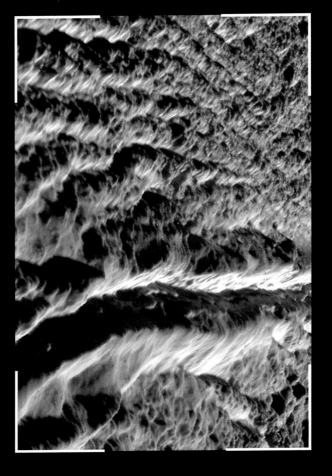

1,433 billion km (890 million miles)

5 km (3 miles)

⊕ TIGER STRIPES

Enhanced-colour images of Enceladus' surface reveal distinct variations in its colour, the most obvious of which are the bluish 'tiger stripes' near the moon's south pole. Here, the tidal heating effects are strongest and the temperature of the icy crust is at its warmest. As a result, the crust remains molten nearly all the way to the surface, and it is here that the geysers erupt most frequently. The four major stripes are known as Damascus, Baghdad, Cairo and Alexandria Sulci.

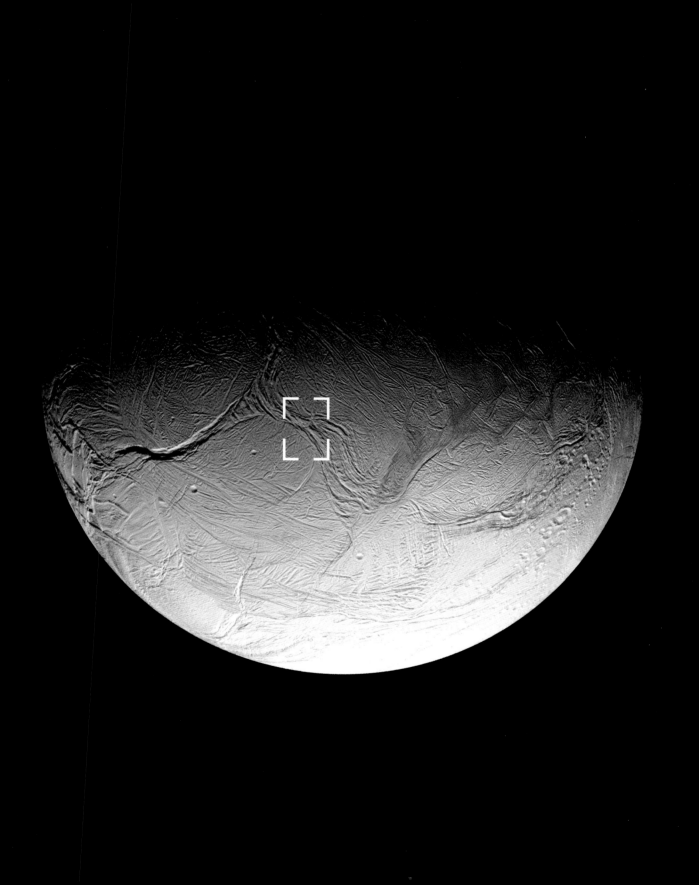

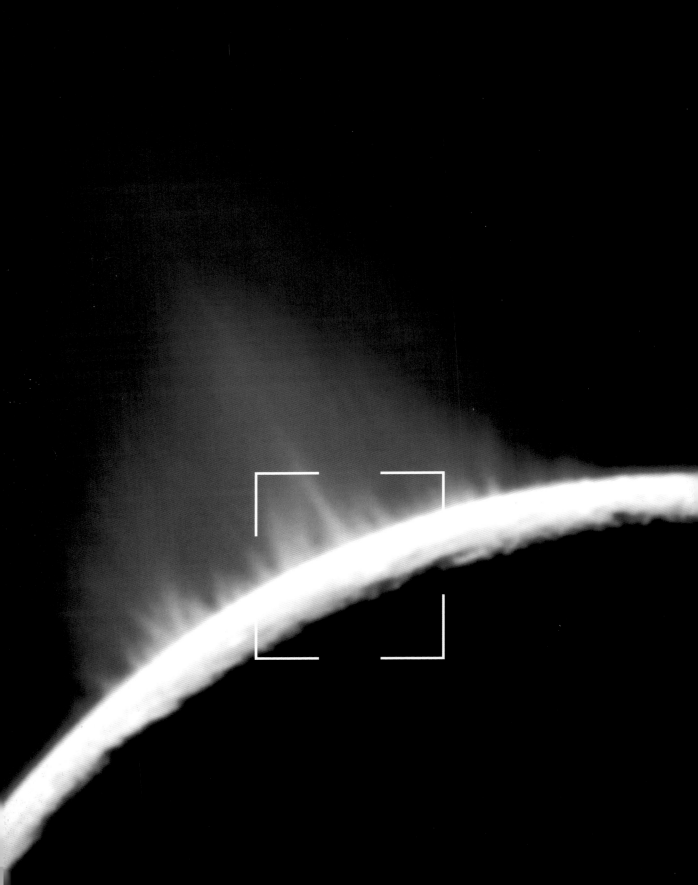

41

ENCELADUS: ICY OUTBURSTS

The existence of snow geysers on Enceladus remained purely hypothetical until NASA's Cassini probe, in orbit around Saturn, accidentally flew through the middle of one during a close encounter with the moon. Backlit images of the moon with the Sun behind it later revealed the full extent of these huge plumes. Measurements from Cassini's sensors showed that the ice crystals are almost pure water, with little sign of the ammonia 'antifreeze' scientists had expected would be present. So it seems that Enceladus' underground reservoirs must be even warmer than expected – heated to perhaps just a few degrees below freezing point on Earth.

GEYSER SOURCES

Geysers on Enceladus, like those on other planets, form when a reservoir of liquid water is 'superheated' to above its natural boiling point, then escapes through cracks in the surface, boiling away into space with spectacular ferocity. Several of the geysers now identified on Enceladus have been traced to parallel ridges associated with the tiger stripe features, known as sulci. Much of the 'snow', formed as escaping water crystallizes in space, falls back to blanket the surrounding landscape, but some escapes completely from the moon's gravity, falling into orbit around Saturn in the tenuous 'E ring'.

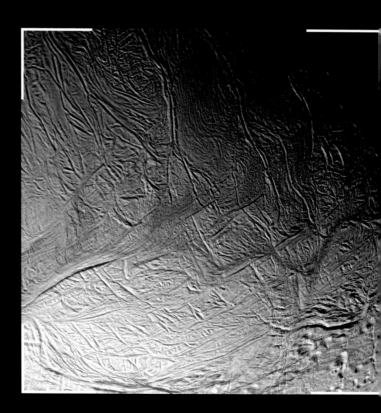

Tethys represents a step up in size from Saturn's innermost moons, and is a near-twin of Dione, the next moon out. As a satellite, it's also remarkable because it shares its orbit with two much smaller worlds – Telesto and Calypso, which circle Saturn precisely 60 degrees in front of and behind Tethys itself. Such 'co-orbital' moons are rare in the solar system, and the handful of other examples consist of pairs of small moons rather than substantial worlds like Tethys.

■ ODYSSEUS CRATER

Tethys's icy, heavily cratered surface is dominated by a huge impact basin called Odysseus. But despite its size, this ring-like feature is fairly shallow – evidence that Tethys's surface is prone to 'slump' over time. Measurements of Tethys's density indicate that it is composed almost entirely of frozen water ice, which can flow very slowly to 'even out' the surface. When mixed with ammonia, this ice can become a fairly runny fluid, capable of erupting through fissures in the surface in a process called 'cryovolcanism'.

∨ ∨   1.433 billion km (890 million miles)

∨ ∨   445 km (276 miles)

Dione is another mid-sized icy moon, similar in size and general appearance to Tethys, although measurements of its density reveal that it must contain considerably more rock. Like its neighbour, Dione shows a history of heavy cratering, especially on its 'leading' hemisphere (that faces forwards as Dione moves around its orbit). There are also traces of cryovolcanic eruptions in its past. This low-temperature equivalent of the volcanism seen on rocky planets is thought to have helped resurface parts of several satellites across the outer solar system.

ICY CLIFFS

Dione's most intriguing feature are the brilliant white streaks of so-called 'wispy terrain' that cover much of its backward-facing 'trailing' hemisphere. Discovered by the Voyager probes in 1980, they were thought at first to be some kind of volcanic fissures that had erupted fresh ice onto the surface. However, in 2004, a Cassini flyby of the moon revealed the dramatic truth – each apparently insubstantial wisp is in reality a towering cliff, up to several hundred metres high, with a reflective face of brilliant ice – traces of huge stresses endured by Dione early in its history.

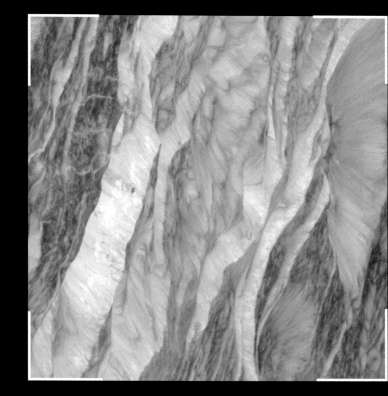

Saturn's second-largest satellite, Rhea, looks at first glance like a larger version of Dione. It, too, has a heavily cratered forward-facing surface and a darker trailing hemisphere streaked with bright wispy terrain that is probably formed from cliffs. However, craters on Rhea are rather better defined than on Dione or Tethys, suggesting that its ice is less prone to flowing and slumping. For some reason, ice on Rhea appears to crystallize in a more rigid form that holds the 'memory' of ancient impacts far better.

■ CRATERED SURFACE

Shortly after Cassini's first flyby of Rhea, scientists analysing its images noticed a faint line of brighter material dotted around the planet's equator. One suggestion was that Rhea may have its own tenuous ring system, with particles occasionally falling to the surface, but attempts to observe these rings have so far proved unsuccessful.

Another important variation in Rhea's surface is a difference in crater sizes across the moon. Some areas have a broad mix of crater sizes, while others contain no craters larger than 40 km (25 miles). This suggests that large parts of Rhea were resurfaced during an early phase of its history, wiping away the larger, earlier craters.

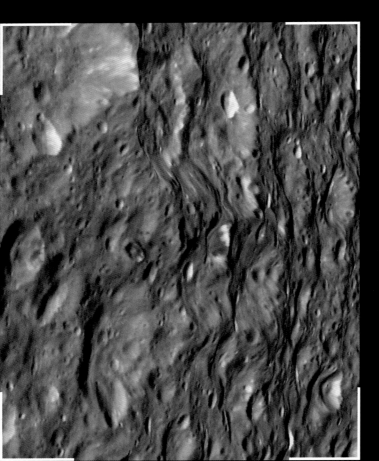

TITAN: MYSTERIOUS MOON

By far the most fascinating of Saturn's moons is Titan, the only satellite in the solar system that is large enough to hold onto a substantial atmosphere (the fractionally larger Ganymede cannot manage this because it receives too much heat from the Sun). Titan is shrouded by a dense orange haze that prevented early space probes from seeing its surface – while the majority of the atmosphere is in fact nitrogen, a small proportion of methane is responsible for both its clouds and its colour.

LIFTING THE VEIL

Titan was an important target for NASA's Cassini orbiter, and so the space probe was fitted out with near-infrared cameras capable of piercing the haze and revealing features on the landscape beneath. The surface of Titan proved to be eerily Earth-like in structure, with obvious 'continents', what look like seabeds, and even river bays and estuaries. The giant moon's landscape has clearly been shaped by fluid erosion, but with temperatures of around −180°C (−292°F), the erosive force is not water, but liquid methane.

TITAN: DESCENT TO THE SURFACE

When the Cassini probe arrived at Saturn in 2004, one of its first acts was to release a small robotic lander built by the European Space Agency. Huygens (named in honour of the Dutch astronomer who discovered Titan's existence) parachuted into the atmosphere of the giant moon on 14 January 2005, and sent back a series of images during its descent and eventual landing. This aerial view of the landing site (opposite) looks very much like a dried-up coastline or lake-shore, with eroded bright highlands rolling down to a dark, flat plain.

■ LANDING SITE

Huygens came to rest in a dried-out 'river delta', surrounded by debris washed down from the highlands. This image of the landing site shows a variety of small rocks and pebbles embedded in darker 'soil'. The rocks, like much of Titan's landscape, are thought to be largely composed of water ice, while the soil is coloured by oily chemicals from the atmosphere. Although the surface where the probe landed was dry and covered by a thin layer of frost, there was strong evidence that it had recently been 'raining', helping to bolster the theory that Titan had a methane cycle, similar to the water cycle that helps shape much of Earth's environment.

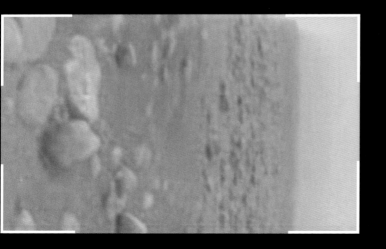

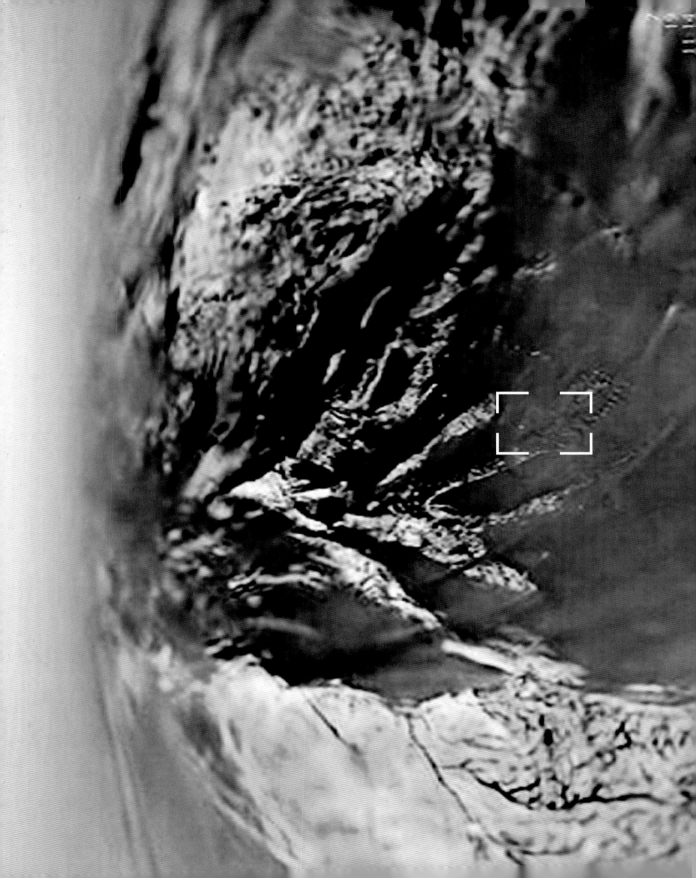

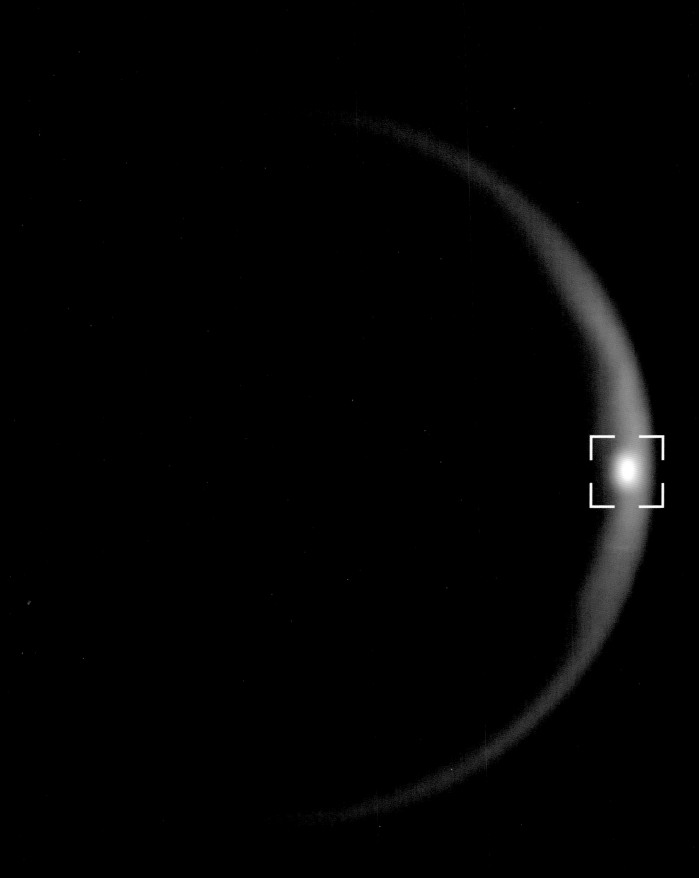

TITAN: METHANE LAKES

This remarkable view from the Cassini orbiter (opposite) captures sunlight glinting off a large lake near Titan's north pole in July 2009. Taken by Cassini's Visible and Infrared Mapping Spectrometer, it offers final confirmation of the presence of liquid methane on the surface of Titan. The reflection has been traced to a large lake or sea called Kraken Mare, previously detected by radar mapping and covering an area of 400,000 sq km (150,000 square miles), at about 71° North.

▦ POLAR VIEW

This radar image of Titan's north polar region shows a landscape strewn with large lakes, most of which are larger than the Great Lakes of North America. Similar lakes have been detected at Titan's south pole, but so far no standing bodies of liquid have been found closer to the equator, despite obvious lakebeds. It seems likely that Titan's climate is delicately balanced, and the distribution of liquid on its surface moves around the satellite, through evaporation and precipitation, throughout its long cycle of seasons.

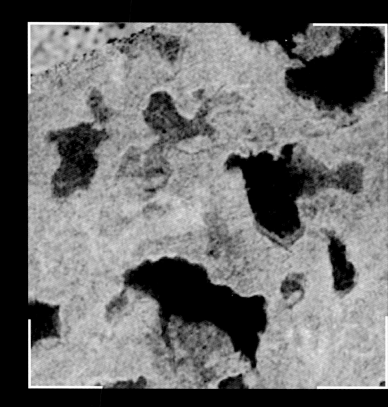

orbiting some way beyond Titan, misshapen Hyperion may deserve the title of strangest moon in the solar system. To begin with, it does not even have a stable rotation period or spin axis – instead it tumbles over and over in a chaotic manner. It's also exceptionally large to be the shape it is – it should have had enough gravity to pull itself together into a spherical shape as it formed. For this reason, it seems likely that Hyperion is just a fragment of a much larger moon that broke apart in some forgotten cosmic collision, perhaps with a large comet or asteroid pulled inwards by Saturn's gravity.

◼ MOTTLED WORLD

The weirdest thing about Hyperion is its appearance. What seems like heavy cratering from a distance resolves itself into a spongy structure on closer inspection. Dark material accumulates in 'pits', while razor-blade ridges are composed of brighter, ice-rich material. False-colour images such as this one from Cassini help to highlight the different materials on the surface. Hyperion's bizarre structure is almost certainly the result of weak sunlight slowly evaporating the icy element of Hyperion's rock/ice surface, causing dark rocky material to crumble inwards, while the remaining lighter material is fashioned into bright peaks.

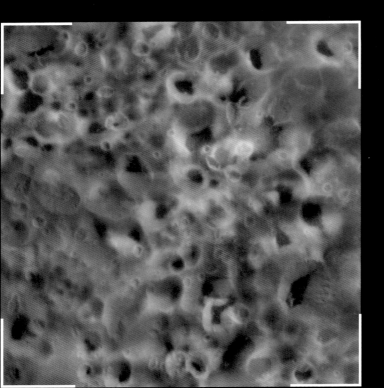

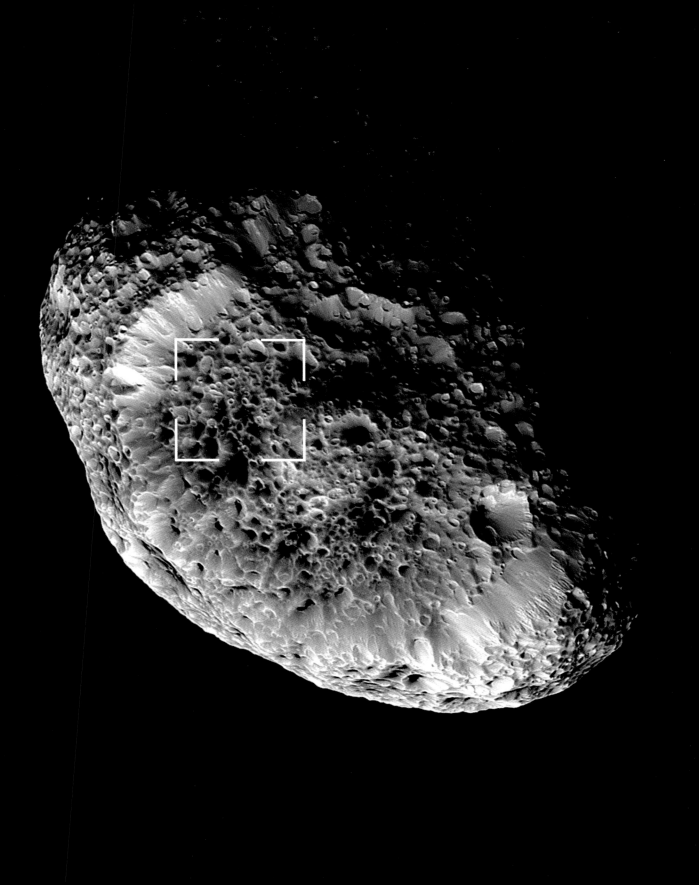

Saturn's third-largest moon, Iapetus, puzzled astronomers from the moment of its discovery in 1671. They found that it was easily visible while on one side of its orbit, but disappeared completely on the other side. Because Iapetus (like Earth's Moon and almost every other satellite in the solar system) is 'tidally locked' with one face permanently facing Saturn, there was only one logical explanation – the satellite's 'leading' hemisphere, which points forward along its orbit, is much darker than its 'trailing' or backward-facing one. When the Voyager space probes first flew through the Saturnian system around 1980, they confirmed this was the case.

LIGHT AND DARK TERRAIN

Cassini images of Iapetus' surface reveal the boundary between light and dark terrains in much more detail. The dark surface seems to overlay the brighter one, and there are no shades of grey between them. The most popular explanation is that the dark material is rocky 'lag' left behind by the slow evaporation of ice from parts of a generally bright surface. Because dark surfaces absorb more heat than bright ones, the process would naturally reinforce itself, but how did it start? The most likely explanation is that Iapetus' leading hemisphere picks up a thin film of dark dust (perhaps originating on Phoebe – see page 118) as it moves along its orbit.

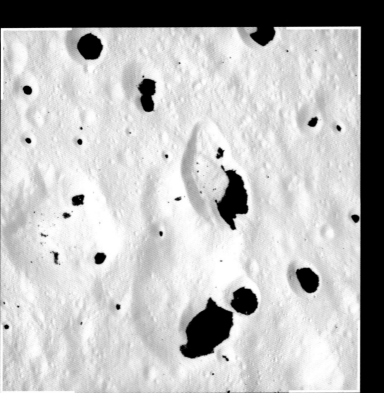

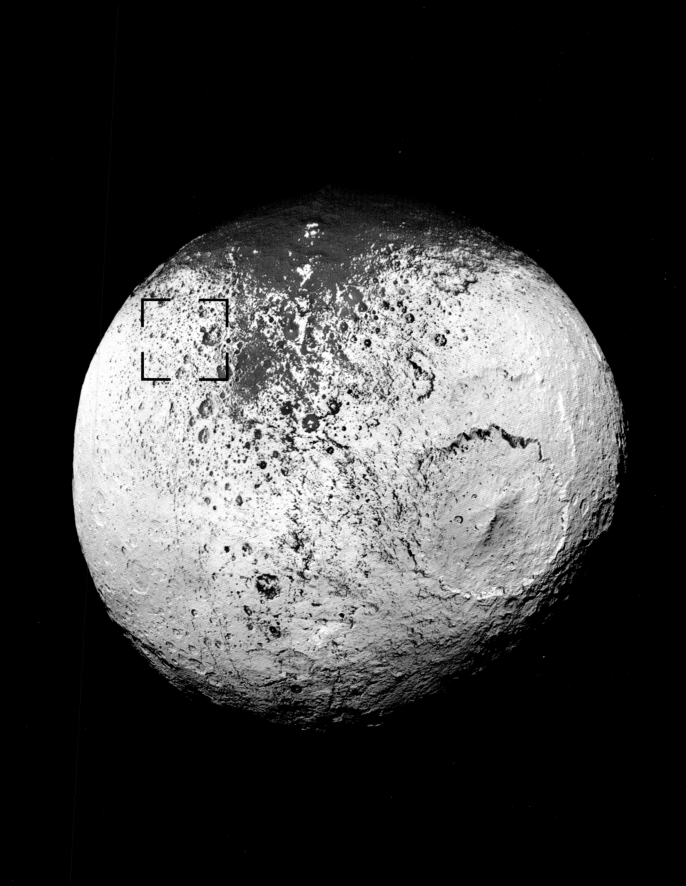

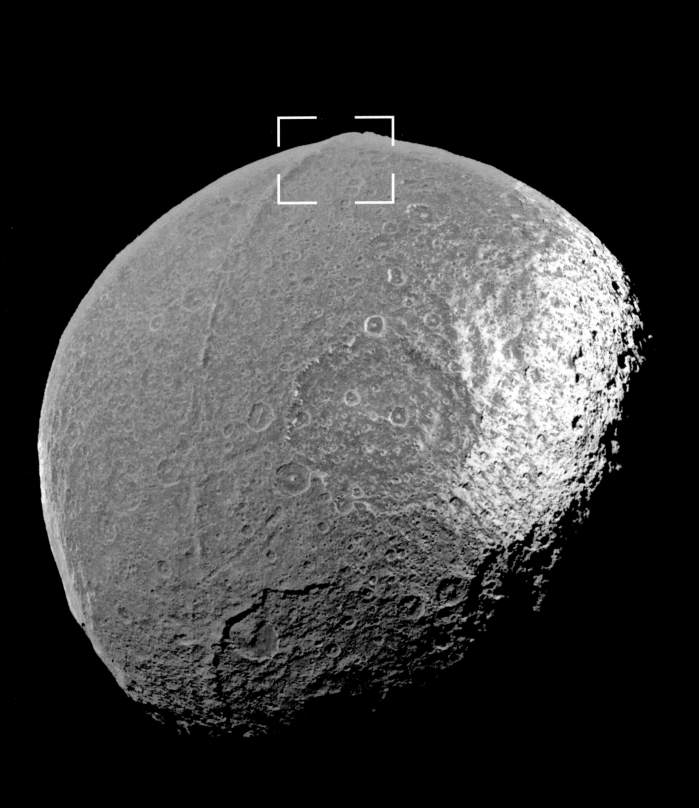

IAPETUS: EQUATORIAL RIDGE

Iapetus' two-tone appearance is not the only strange thing about it – in 2004, NASA's Cassini probe discovered that the moon is encircled by a ridge that runs in a straight line around much of the equator, giving it an overall shape similar to a walnut. Astronomers are still arguing over the cause of this remarkable feature. One idea is that the ridge is a 'fossil' from a time when the moon rotated much faster and had a pronounced equatorial bulge. Another is that it originated as an outburst of ice from beneath the crust, and that gravitational forces from Saturn changed the satellite's orientation to minimize the tidal effect of its unusual shape.

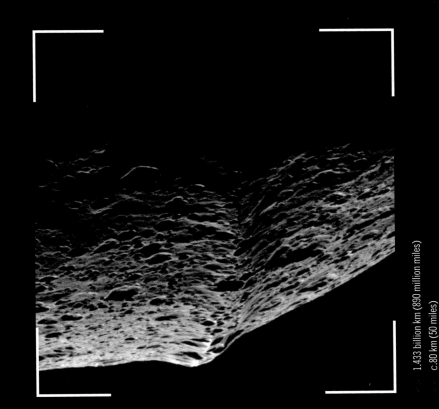

AROUND THE EQUATOR

Iapetus' equatorial ridge is best defined where it crosses the large area of dark terrain known as Cassini Regio. The central section is some 1,300 km (800 miles) long, 20 km (12 miles) wide and up to 20 km (12 miles) high in places. Elsewhere, the ridge is broken into short outcrops and a series of isolated mountains up to 10 km (6 miles) high that are dotted along the equator in the brighter regions of the surface.

1.433 billion km (890 million miles)

c.80 km (50 miles)

Beyond Iapetus, Saturn (like all the giant planets) is orbited by a swarm of 'irregular satellites' – captured asteroids and comets in long elliptical orbits. The innermost of these, Phoebe, is also by far the largest. For this reason, it was specially targeted for a flyby as the Cassini probe flew into orbit around Saturn in 2004. The first images revealed a roughly spherical world with a diameter of about 220 km (140 miles), covered in craters. Phoebe's surface is also extremely dark – quite literally pitch-black – and a stark contrast to the bright landscapes of most of Saturn's inner satellites.

■ BATTERED WANDERER

Close-up images of Phoebe leave little doubt that this strange satellite is a large comet-like object, captured into orbit around Saturn long ago. Initially, its dark surface led to the assumption that it was an asteroid, but further analysis revealed that many of its craters have comparatively bright floors. Phoebe's dark crust seems to lie on top of a largely icy body, and astronomers now think the satellite may be our first glimpse of a 'Centaur' – one of a class of small icy worlds that follow independent orbits around the Sun in the region between Saturn and Neptune.

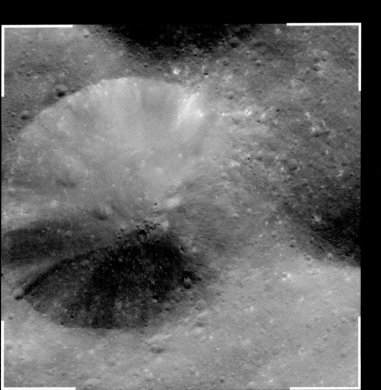

[52] URANUS

The seventh planet from the Sun, Uranus was also the first to be discovered in recorded history – by astronomer William Herschel in 1781. At roughly twice the distance of Saturn and about half its size, it was little more than a blur through ground-based telescopes until the 1980s, and our first detailed look at it came from Voyager 2's brief flyby in 1986. Voyager images (opposite) revealed a placid, featureless blue-green world. Its most intriguing feature seemed to be its extreme axial tilt – Uranus orbits the Sun 'tipped over', so that one pole and then the other faces the Sun during each 84-year orbit.

■ GATHERING STORMS

The launch of the Hubble Space Telescope and improvements to ground-based instruments mean that we can now study Uranus from Earth, and images taken since the 1990s have shown a very different world from that seen by Voyager. As Uranus moves from northern winter to a phase in which different parts of the planet receive more even illumination, it has developed distinctive bands and bright storms in its atmosphere. It seems that the planet's extreme seasons help to suppress the normal weather systems around each midsummer and midwinter.

<> 2.88 billion km (1.79 billion miles)

>< 51,118 km (31,750 miles)

the rings around Uranus were discovered by accident in 1977, when astronomers waiting to record the planet's passage in front of a star recorded a series of dips in its brightness before and after the main 'occultation' event. We now know that Uranus has 13 distinct rings – very thin, dark and tightly defined compared to the enormous planes that encircle Saturn. Seen from Earth, these rings immediately give away the planet's strange orientation, and from the right angle can give Uranus a 'bullseye' appearance.

◼ VOYAGER'S VIEW

Voyager 2 snapped this detailed image of the major rings during its 1986 flyby. The brightest and outermost ring is known as the Epsilon Ring. While Saturn's ring particles are composed of highly reflective water ice, those orbiting Uranus are rich in methane ice, which is far less reflective. Methane is an important chemical in this part of the solar system – its tendency to absorb red and orange sunlight creates the distinctive blue-green colours seen in the atmospheres of both Uranus and Neptune.

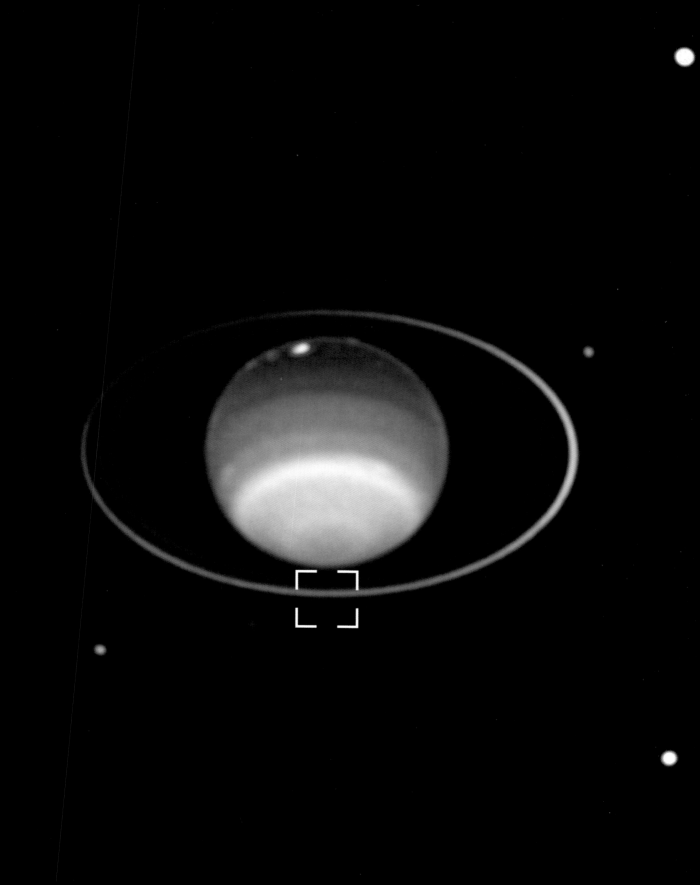

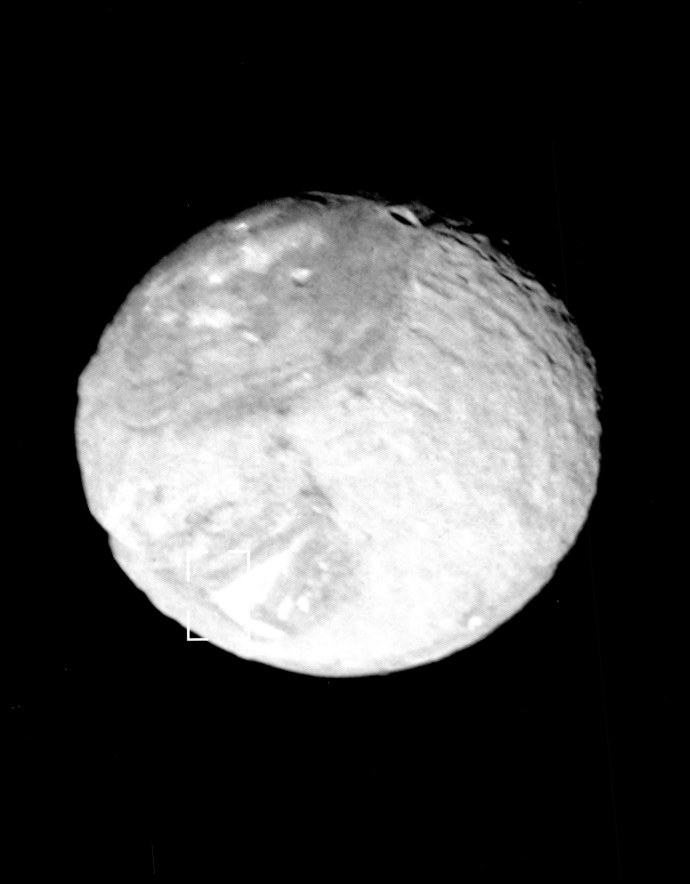

MIRANDA

Uranus has a system of at least 27 satellites, ranging from small 'shepherd moons' in and among the rings, to captured comets in irregular orbits. In between there are five substantial, more or less spherical worlds – Miranda, Ariel, Umbriel, Titania and Oberon. Miranda is the smallest and closest to Uranus, but also the most intriguing. Its surface is a complex mishmash of countless different types of terrain, formed at different times, and by different processes. When Voyager sent back its first image of this Frankenstein moon, many astronomers assumed it had been broken apart by an ancient collision, and reassembled itself in its current state.

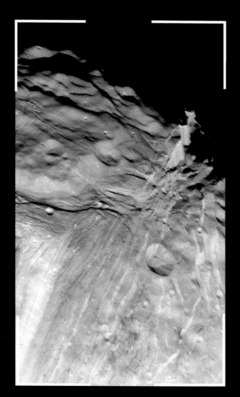

▷◁ 2.88 billion km (1.79 billion miles)
>< 60 km (37 miles)

■ JUMBLED WORLD

Verona Rupes in Miranda's southern hemisphere are some of the highest cliffs in the solar system, towering up to 5 km (3 miles) high. Studies of other features on the moon's surface suggest a long history of complex geological activity including faulting of the surface and the eruption of 'cryovolcanoes' exuding a runny 'lava' of rock, ice and ammonia. This activity was probably caused by tidal heating at a time when Miranda's orbit was more eccentric than it is today – the molten state of the moon's interior allowed the surface to separate and rearrange itself without the need for Miranda to break apart completely.

NEPTUNE: PLANET OF STORMS

When Voyager 2 flew past deep blue Neptune, the most distant major planet, in 1989, astronomers found an unexpectedly stormy world. They had expected Neptune would be as apparently placid as its inner neighbour Uranus, and they did not yet suspect that Uranus' featureless appearance was itself just a passing phase. Both Uranus and Neptune are 'ice giants', composed of a mix of hydrogen gas and other more complex chemicals – such as methane, water and ammonia – that form slushy 'ices' beneath their outer atmospheres. Neptune's more intense colour is due to a larger quantity of red light-absorbing methane in its atmosphere.

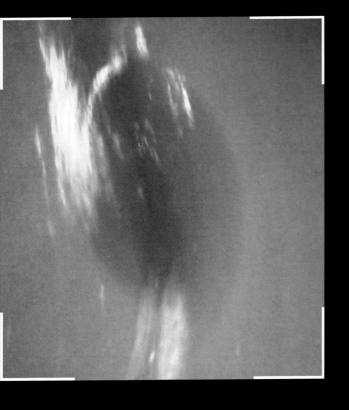

■ THE EYE OF NEPTUNE

The Great Dark Spot (GDS) was a huge storm seen raging in Neptune's atmosphere during Voyager 2's 1989 flyby (opposite). But despite its apparent similarity to Jupiter's Great Red Spot, the GDS seems physically quite different – while the Red Spot is an anticyclonic weather system capped in colourful clouds, the Dark Spot appears to be a 'hole' through the upper layers of the cloud deck. The GDS also proved to be nowhere near as long-lived as Jupiter's giant storm – by the time the Hubble Space Telescope turned its gaze to Neptune in the mid-1990s, it had faded from view, although other storms of equal magnitude have since appeared.

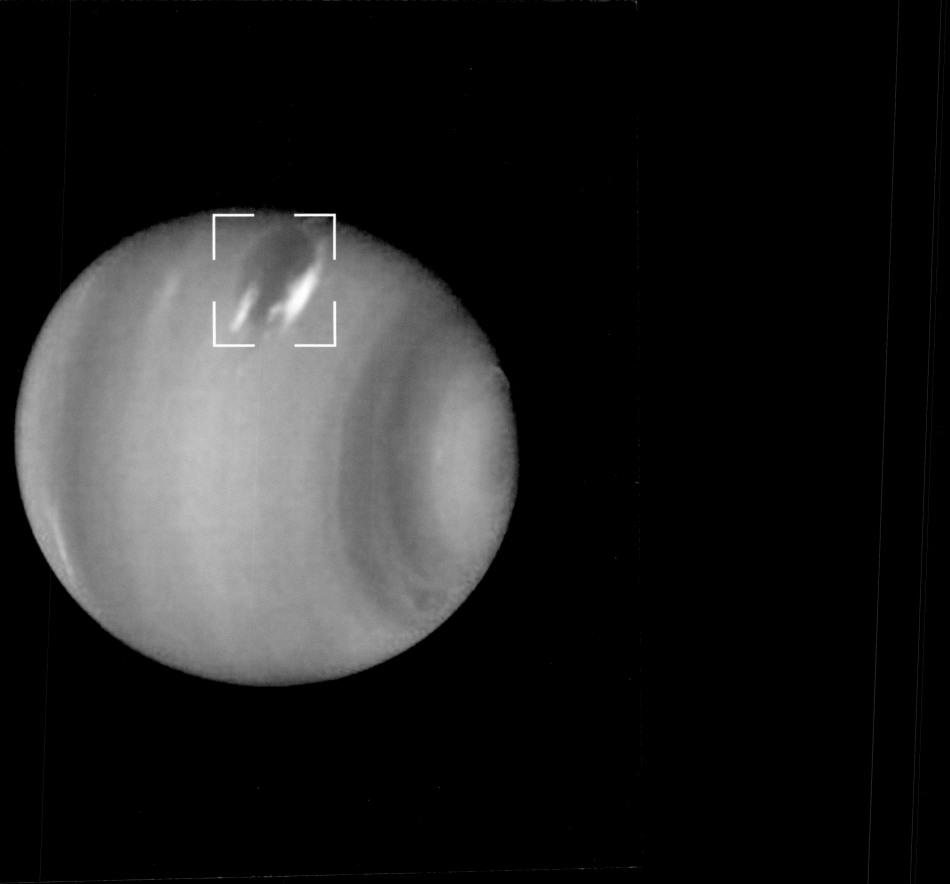

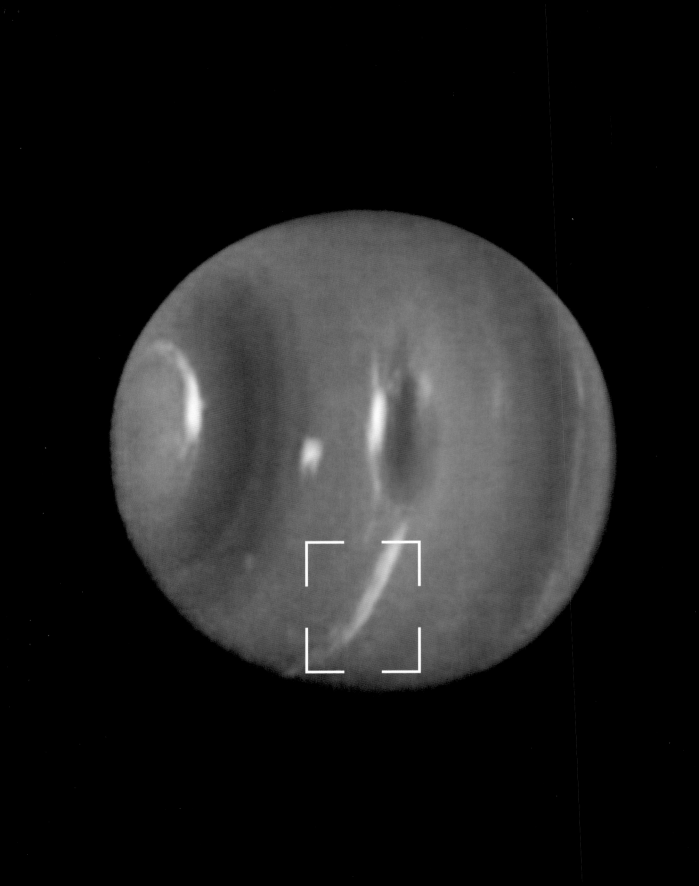

56

NEPTUNE: HIGH-SPEED CLOUDS

Neptune is home to some of the highest winds in the solar system, ripping around the planet at up to 600 m/s (1,300 miles per hour) – almost supersonic speeds. These winds propel small white clouds known as scooters around the planet, and also wrap narrow bands of high-altitude wispy cloud around Neptune's circumference. This far from the Sun, the heat provided by sunlight is far too weak to power such energetic weather systems – instead it seems that Neptune, like Jupiter and Saturn but (so far as is known) unlike Uranus, generates energy deep within itself.

■ POWERED FROM WITHIN

Because the interiors of the gas giant planets are fluid, the motion of masses of gas or liquid as they pass each other generates friction and releases heat that gradually makes its way to the surface, powering dynamic weather systems even this far out in the Sun's heat. This internal energy output can be boosted by changes to the atmosphere's internal chemistry and, in Neptune's case, the chemistry involved may be remarkable. It's thought that in a layer deep beneath the surface, carbon atoms from compressed liquefied methane are forced together to form minute crystals of diamond, releasing energy in the process.

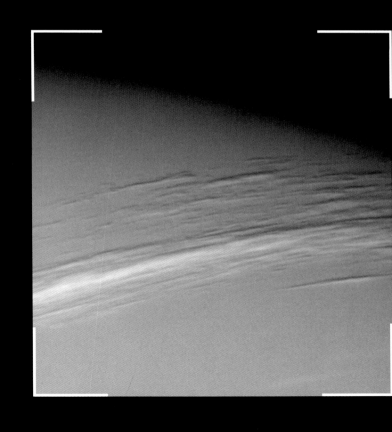

Neptune has a family of 13 known moons, and stands out among the giant planets because only one of these is a substantial world – all the remainder are either small inner shepherds around Neptune's tenuous ring system, or distant irregular satellites. Triton, however, is unusual in other ways – for one thing it orbits Neptune in the 'wrong' direction relative to the planet's rotation, and for another, its orbit is perfectly circular. These are both signs that Triton is an interloper – a rogue 'dwarf planet' swept up in Neptune's gravity, whose sudden arrival probably disrupted Neptune's original family of moons.

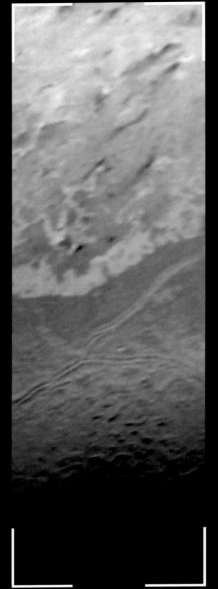

∧ >   4.5 billion km (2.8 billion miles)

∨ <   c. 800 km (500 miles)

■ CANTALOUPE TERRAIN

As Triton settled into its current orbit, it would have suffered extreme tides that would have heated the captured moon, melting it from the inside out. The most obvious sign of this is the pitted 'cantaloupe terrain' that covers much of Triton's surface (on the right in these images). It's believed to have formed when warm 'diapirs' of ice forced their way up through colder surroundings, rather like bubbles rising through a glass of soda.

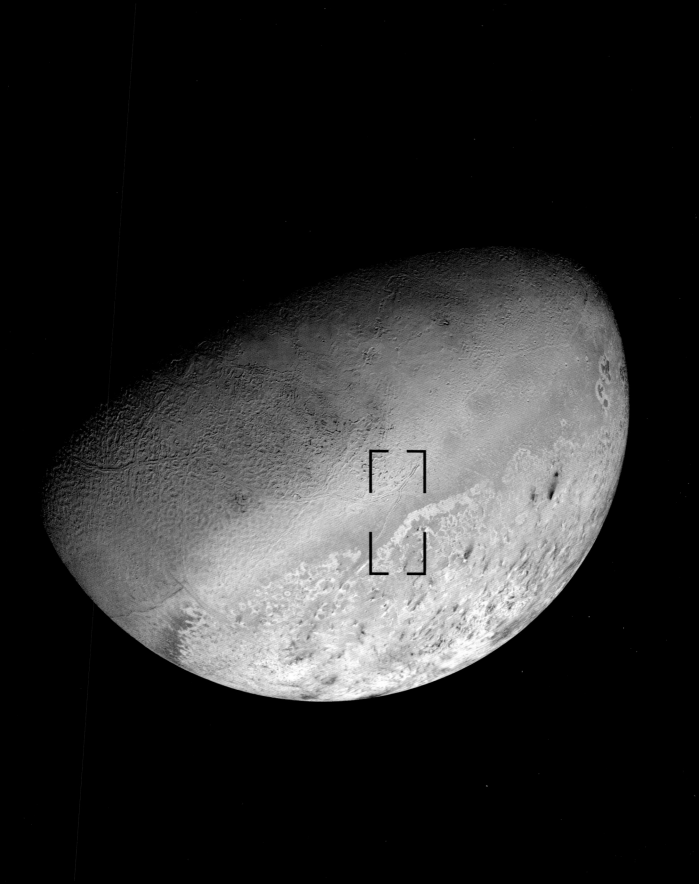

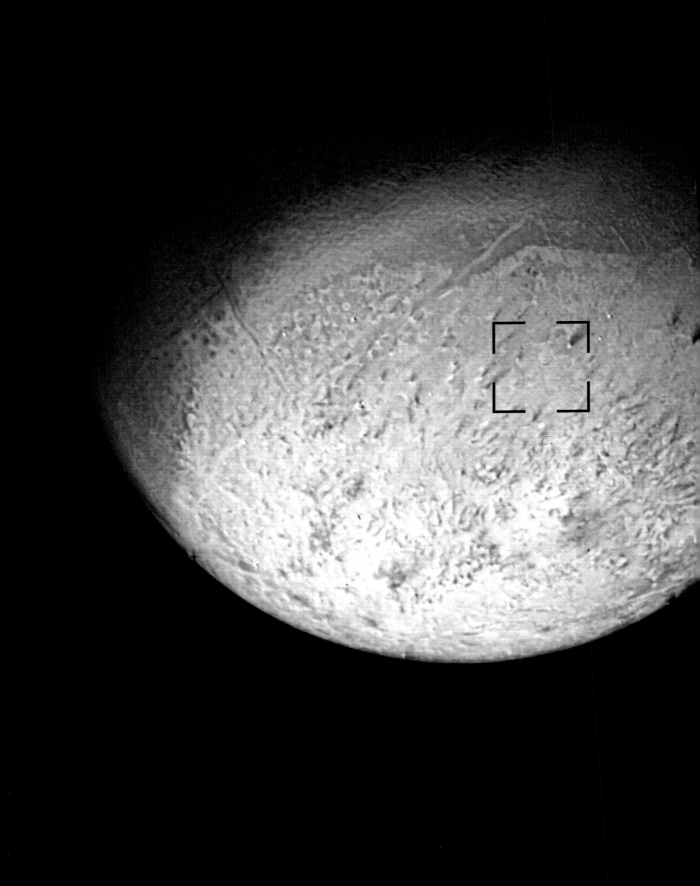

TRITON: ICE GEYSERS

Perhaps the most surprising aspect of Triton is that, despite average surface temperatures of −235°C (−391°F), the moon's surface is geologically active. While much of the surface consists of the hummocky cantaloupe terrain, some much smoother areas are streaked with dark lines of soot leading away from fissures in the surface. When photographed from shallow angles, these streaks proved to lie beneath long geyser-like plumes of gas and soot billowing out from the surface. It seems that Trton's interior is still warm enough to melt ices beneath the surface so that they escape violently through the surface.

■ GEYSER STREAKS

The spectacular ice geysers reveal the presence of a thin, otherwise invisible nitrogen atmosphere around Triton.

Plumes of soot-laden vapour rise straight into the sky before being swept up in high-altitude winds. These winds seem to be moving from Triton's summer pole to its winter pole, suggesting a simple weather system that probably reverses over time. (Heating from the Sun also seems to play an important role in triggering the eruptions in the first place.) As dust falls back out of the atmosphere, it leaves streaks for up to 150 km (93 miles) downwind of the geyser itself.

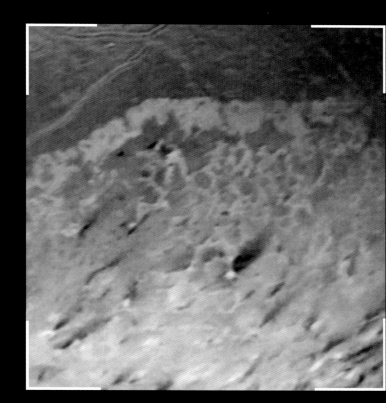

Orbiting the Sun on an elliptical path that actually comes within Neptune's orbit for 20 Earth years out of each 248-year circuit, Pluto is a tiny world that was classified as the ninth true planet until 2006. It is now officially a dwarf planet, among the largest members of a doughnut-shaped belt of small icy worlds beyond Pluto known as the Kuiper Belt. Despite its small size, Pluto has three satellites. The largest, Charon, is half the size of Pluto itself, while the much smaller Nix and Hydra were only discovered by astronomers using the Hubble Space Telescope in 2005.

■ SURFACE DETAIL

From Earth, Pluto appears as a tiny point of light through even the largest telescope, but astronomers can study the way in which its brightness and colour vary as it rotates and occasionally disappears behind Charon. The maps produced in this way show that Pluto has a varied surface, with bright patches of ice that may partially evaporate when it is closest to the Sun, producing a thin atmosphere. Our limited knowledge of Pluto should receive a big boost when NASA's New Horizons probe flies past in 2015.

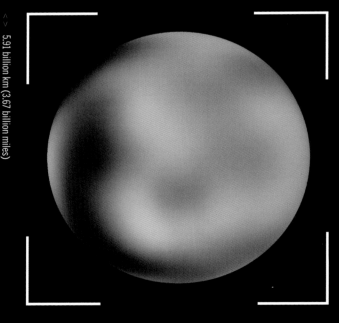

< > 5.91 billion km (3.67 billion miles)
> < 2,306 km (1,432 miles)

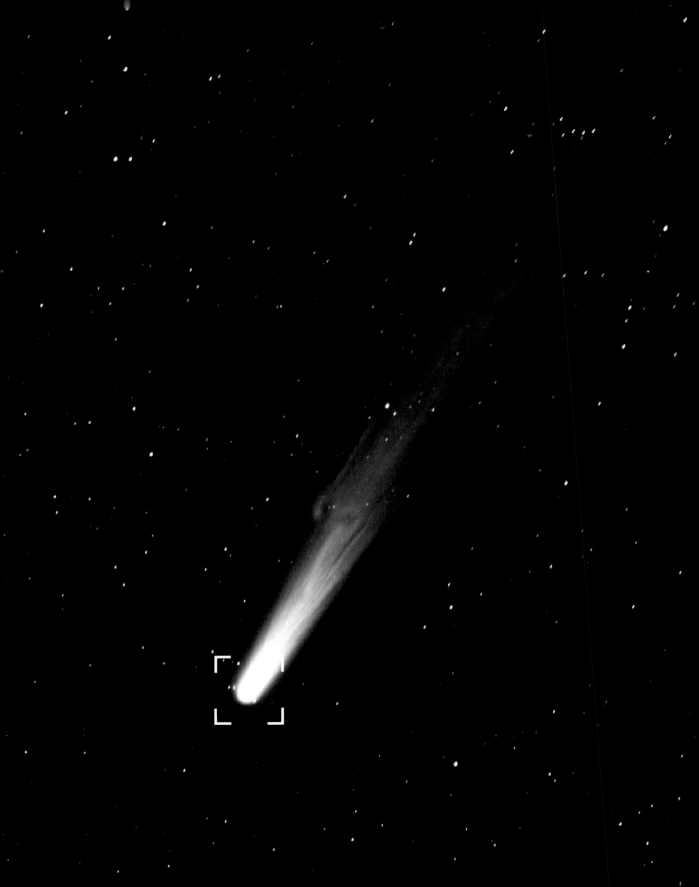

HALLEY'S COMET

Comets are small icy bodies that spend most of their time among the dwarf world of the Kuiper Belt, or even further from the Sun in a vast spherical shell, two light years across, known as the Oort Cloud. They only become visible when they fall into orbits that bring them past the Sun, at which time their dark surfaces attract heat that melts the ice beneath. Plumes of gas and water vapour erupt from beneath the surface to envelop the comet's solid 'nucleus' in a planet-sized atmosphere called a coma, which may be caught up in the solar wind of particles and radiation streaming out from the Sun, creating a long tail behind the comet.

■ HALLEY UP CLOSE

Halley's Comet, shown in these images, is the most famous and best-studied comet of all. It is a short-period comet, having fallen into an orbit that brings it back past Earth and the Sun every 76 years, and has been recorded by astronomers since at least 240 BC. During its 1986 passage around the Sun, Halley was targeted by an international armada of space probes, including the European Giotto probe, which flew within 600 km (370 miles) of the 16x8-km (10x5-mile) nucleus and returned images of its dark, active surface.

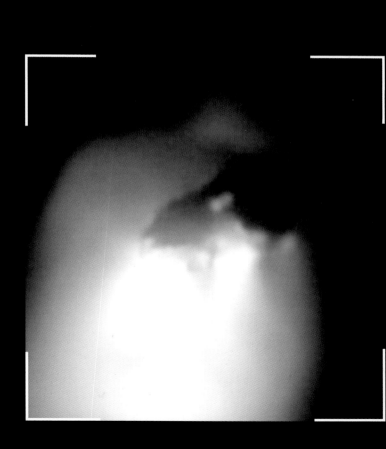

In 2003, astronomers looking for new 'Kuiper Belt Objects' (KBOs) in the far reaches of the solar system finally discovered what they had long expected – a world larger than Pluto. 2003 UB313, now known formally as Eris after the Greek goddess of discord, was hailed at the time as a 'tenth planet', but it is only around 6 per cent larger than Pluto, and the debate about its status led to the invention of the new 'dwarf planet' category for independent worlds, such as Pluto and Eris, which are large enough to pull themselves into a spherical shape, but not large enough to 'clear out' the neighbourhood of their orbits.

■ MYSTERY WORLD

With a long 557-year orbit that takes it out to three times the distance of Neptune, Eris is even more difficult to study than Pluto, but astronomers using the best Earth-based and space-based telescopes have found out a surprising amount. The surface is distinctly grey compared to Pluto and Triton, as shown in this artist's impression, and probably covered with large areas of methane ice. Eris also has a satellite, named Dysnomia, orbiting it in 15.8 days. However, this frozen wanderer remains little more than a blur of pixels through even the most powerful telescope.

10.12 billion km (6.29 billion miles)

OUR GALAXY

All the stars in Earth's skies are members of a huge spiral system called the Milky Way, containing around 200 billion stars in total, alongside vast amounts of interstellar gas and dust. Individual stars are so far away and relatively small that it's almost impossible for us to see them in detail, but they are often associated with larger structures – star clusters, star-forming nebulae, and clouds of discarded gas expelled at various points in their long and varied lives.

Looking up into the night sky can reveal a surprising amount about the nature of stars. Even a brief glance will reveal that they display a wide range of brightness and a variety of colours ranging from blue to red, and are spread unevenly around the sky – densely packed in some places and sparsely scattered in others.

The variation in brightness shows that the stars are either at vastly different distances from Earth, or have intrinsic differences in their true brightness or 'luminosity' – and in fact both are true. The visual brightness of a celestial object seen from Earth is known as its 'apparent magnitude' – the brighter the object, the lower the magnitude. The brightest star in the sky, Sirius, has a magnitude of −1.4, while the faintest stars visible to the naked eye are around magnitude 6. However, Sirius is an unremarkable

white star (roughly 25 times more luminous than the Sun) that just happens to appear brilliant because it is one of the closest stars to Earth, just 8.6 light years away. In contrast, the second brightest star, Canopus, is a brilliant white 'giant' 300 light years away and 14,000 times more luminous than the Sun, and there are countless luminous stars which lie so far away that they cannot even be seen with the naked eye.

Differences in stellar colours, meanwhile, are closely linked to the surface temperatures of stars. Red light has longer wavelengths and lower energies than bluer colours, and so can be emitted by stars with lower temperatures. Of course, the overall colour of any star is formed by a combination of many different wavelengths and subtly different colours of light, but in general there is a clear link

between surface temperature and colour – the reddest stars have temperatures around 3,000°C (5,400°F), then temperatures climb through orange, yellow and white stars to blue-white and blue stars with searing surface temperatures around 10,000°C (18,000°F).

The link between a star's colour and its luminosity can reveal other stellar properties, and astronomers often compare the two on a chart known as a 'Hertzsprung–Russell (H–R) diagram', independently invented by astronomers Ejnar Hertzsprung and Henry Norris Russell around 1910. Because a star's temperature depends on the amount of energy it is releasing and the surface area through which that energy can escape, it is fairly easy to work out a star's relative size – a highly luminous star that is still red must have a truly enormous surface area – (a 'red giant'), while a faint star that is still blue- or white-hot must be absolutely tiny (a 'white dwarf'). In general,

though, most stars seem to follow a pattern that links small, faint, red stars to large, brilliant, blue ones. This dominant group of stars, which appears as a diagonal strip across the H–R diagram, is known as the 'main sequence', and independent measurements of the mass of main-sequence stars (based on observations of their orbits in double or multiple-star systems) shows that their brightness is also related to their mass – the faintest 'red dwarfs' have masses as little as 8 per cent of the Sun's, while the most brilliant 'blue giants' can have up to 50 times solar mass.

One major challenge for astronomers is that they can only study a single 'snapshot' of the sky – a brief moment in the long lives of stars. However, the predominance of main-sequence stars suggests that the vast majority of stars spend nearly all of their lives following the main-sequence 'rule'. Because stars nearly always keep the

26,000 light years (to core)
100,000 light years

OUTSIDER'S VIEW
This artist's impression shows our spiral galaxy, the Milky Way, as it might be seen by an observer hanging above the outer arm and looking past the region of our solar system towards the star-packed core.

Pinkish star-forming nebulae and brilliant open clusters highlight the structure of the spiral arms, but the vast majority of stars, residing in the disc between the arms, are more sedate and similar to our own Sun.

same mass throughout their lives, they tend to stay in the same position on the H–R diagram throughout their long main-sequence lifetime – the Sun, for instance, is a 'yellow dwarf' star roughly halfway through its 10-billion-year life on the main sequence.

Careful study can reveal further important patterns among the stars – for instance, the heaviest and brightest blue stars are only ever found in close proximity to the clouds of gas from which they were born, suggesting that such stars are short-lived and do not survive long enough to drift away from their birthplace. In contrast, the only stars found in globular clusters and other environments that are completely devoid of star-forming gas are predominantly low-mass and red or yellow in colour, suggesting that these are the longest-lived stars of all. By splitting the light of individual stars into rainbow-like spectra, astronomers can also identify the chemical 'fingerprints' of individual elements in stellar atmospheres – dark 'absorption lines' caused by chemicals absorbing light of very specific wavelengths. This reveals that stars are predominantly composed of the lightest elements in the Universe, hydrogen and helium, with only small amounts of the heavier elements termed 'metals'.

Putting all these clues and patterns together to form a coherent picture of stellar evolution was one of the greatest achievements of 20th-century astronomy, and astronomers were helped enormously by our developing understanding of subatomic physics on Earth. Today we know that stars are fuelled by nuclear fusion – the forcing together of atomic nuclei to create heavier elements at

high temperatures and intense pressures only achieved in the heart of stars. This process involves a very small, slow mass loss, which is directly converted into energy in accordance with Einstein's famous equation $E=mc^2$, and is most efficient in converting the lightest atomic nuclei of all (those of hydrogen) into the next lightest, helium. It is this process that powers all stars throughout their main-sequence lifetime, but it can proceed at vastly different rates depending on conditions in the core that are ultimately governed by the star's mass – hence why the most massive stars live shorter lives despite their greater fuel supplies.

It is only when their core supply of hydrogen runs out near the end of a star's life that it must look elsewhere for additional fuel. At first it plunders hydrogen in its outer layers – a process that causes it to brighten enormously but also increase hugely in size to become a red giant. Then, as the burnt-out core collapses, conditions become so extreme that the star can begin to generate energy by fusion of heavier elements. The most massive stars can sustain several cycles of fusion before they are completely exhausted, but the end result of star-death is always essentially the same – the star's outer layers are expelled (in the gentle puffs of a planetary nebula or the violent explosion of a supernova), and the remaining burnt-out core collapses inwards to form either a white-dwarf or an even denser neutron star or black hole. The material scattered across the galaxy by the processes of star-death will go on to be incorporated in new generations of stars and planets.

The vast majority of our galaxy's material is hidden from our everyday view. Some of this missing material is so-called 'dark matter', invisible to telescopes of all sorts by definition (see page 295), but more is simply too cold and dark to glow in visible wavelengths. These huge clouds of gas and dust, often called 'dark nebulae', are normally visible only when they are silhouetted against brighter backgrounds of star clouds or glowing gas. However, infrared telescopes can reveal these clouds in all their glory, such as the complex nebulosity that fills much of the constellation of Orion.

■ HORSEHEAD NEBULA

Orion's Horsehead is probably the most famous of all dark nebulae – instantly identifiable because of the chesspiece-like silhouette it makes against the glowing background nebula IC 434. The nebula is roughly 1,500 light years from Earth, and marks a misshapen pillar of gas and dust, roughly two light years across at the 'head', in the process of collapsing into stars. Its distinctive shape was first noticed by US astronomer Williamina Fleming in 1888. IC 434, meanwhile, displays shimmering curtains of light sculpted by the galaxy's own weak magnetic field.

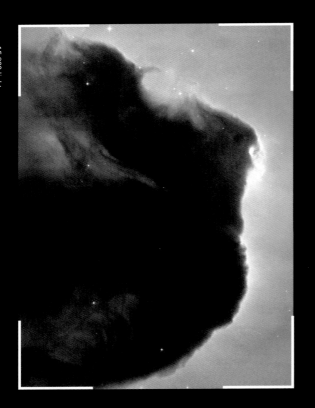

15,000 light years

2 light years

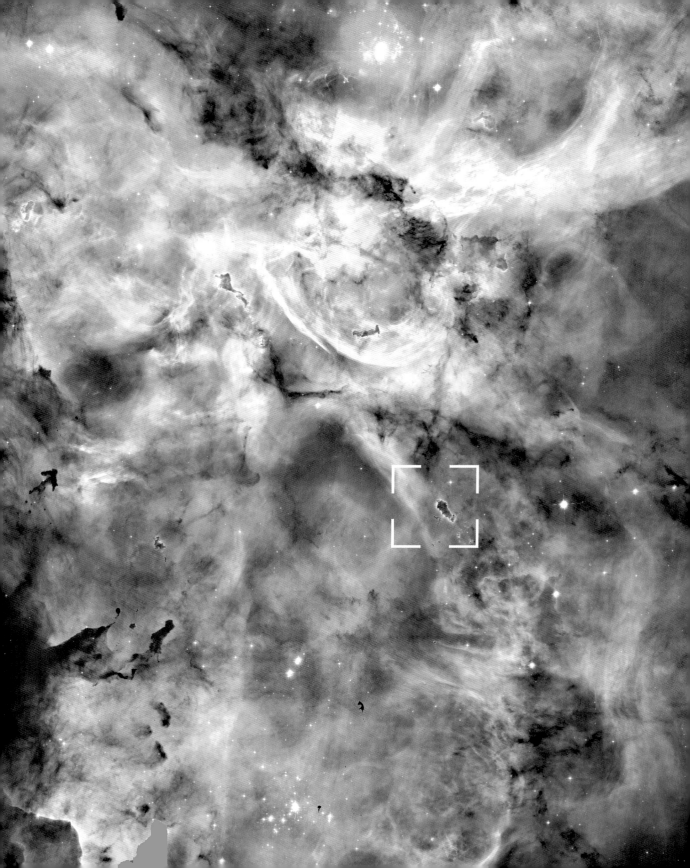

The so-called 'Caterpillar' is a relatively small, elongated dark cloud silhouetted against the nebula's background light. It is roughly one light year long and contains one or more stars in the process of formation. The Caterpillar glows around its edges thanks to a process called photoionization – high-energy radiation from stars in the surrounding nebula is exciting the outer layer of gas and dust so much that individual atoms split apart and become glowing electrically charged particles.

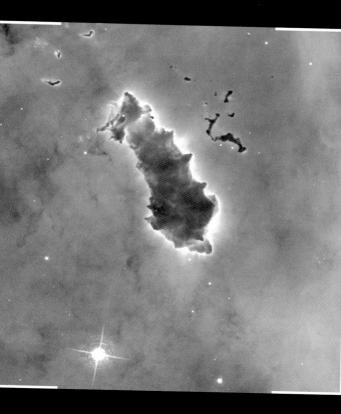

MYSTIC MOUNTAIN

This detailed Hubble image shows a tottering pillar of star-forming gas and dust, eroded into a mountainous top by the forces of ultraviolet radiation blowing out from massive newborn stars nearby.

The Carina Nebula is thought to contain up to a dozen of these stellar monsters with masses around 50 to 100 times that of the Sun, as well as countless smaller stars. The lighthouse-like jets of material emerging near the top of the pillar are caused by 'bipolar outflow' (see page 173) – a sign that star formation is still going on deep within the opaque pillars.

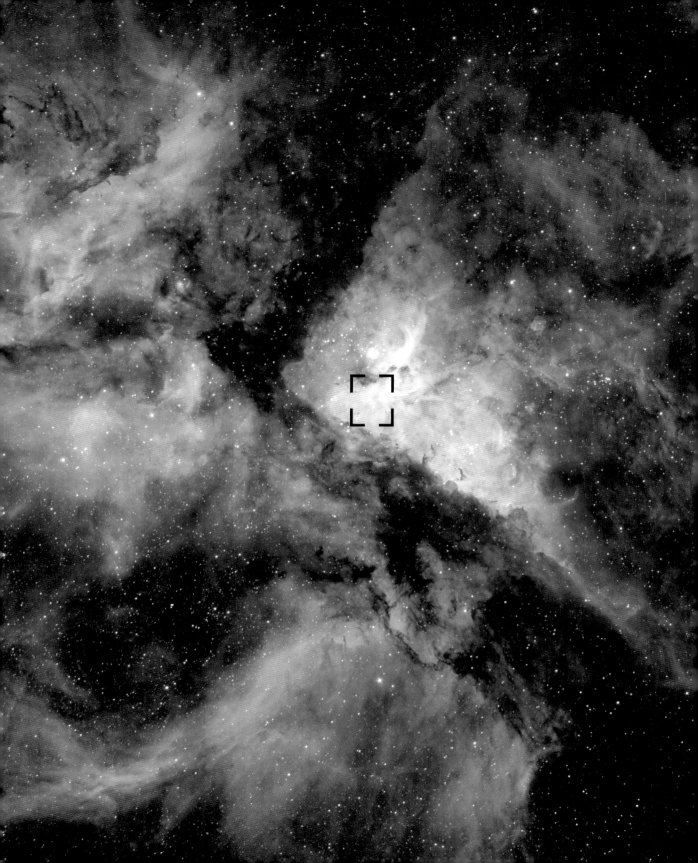

This panoramic view of the Carina Nebula (opposite) was produced at Cerro Tololo Inter-American Observatory in Chile, and combines light from three different elemental processes to highlight the nebula's chemical composition. Red tones indicate the presence of sulphur within the huge interstellar cloud, while green shows the densest areas of hydrogen and blue shows oxyge The colours are also roughly representative of temperatures, with red indicating the cooler regions and blue the hottest.

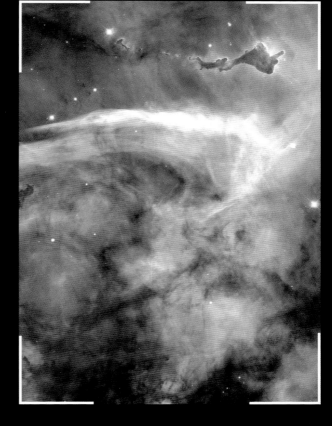

7,000 light years

7 light years

KEYHOLE NEBULA

One of the most distinctive features in the Carina Nebula is the so-called 'Keyhole' – a darker cloud of gas silhouetted against the background glow. The upper part of the Keyhole, roughly seven light years across, is captured in detail by this Hubble Space Telescope image, which reveals its form in a swirl of overlapping clouds of dark dust and glowing gas. The glowing edges to some of the clouds are caused by the action of ultraviolet radiation from a massive star just out of the frame.

Perhaps the most beautiful of all the sky's nebulae is the Great Orion Nebula, M42. Visible from both the northern and southern hemispheres, it is easily identified as a rosette-like knot of light within the 'sword' of the constellation Orion (the Hunter). The nebula is just one small part of the enormous 'Orion molecular complex' – a mostly dark cloud that stretches beyond Orion and its neighbouring constellations. Its flower-like structure is illuminated by stars forming at its centre, including the concentrated cluster known as the Trapezium (see page 177).

■ CHEMICAL COLOURS

A combined ultraviolet, infrared and visible-light image from the Spitzer and Hubble Space Telescopes reveals something of the Orion Nebula's true complexity. Ultraviolet regions shown in green mark clouds of hydrogen and sulphur heated by the radiation from the central stars, while orange and yellow blobs show young stars still embedded in their cocoons of gas and dust and only visible at infrared wavelengths.

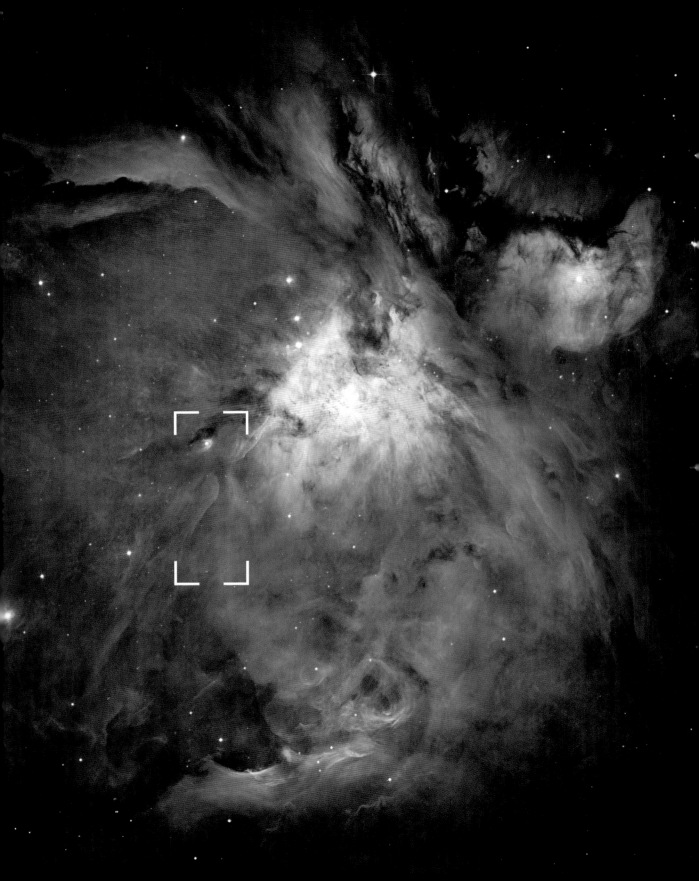

ORION NEBULA 2

This detailed Hubble Space Telescope view of the Orion Nebula (opposite) reveals more than 3,000 stars in visible light. Most of them lie within or close to the nebula itself – only a handful are positioned in the foreground from our point of view. As fierce radiation from the largest of these stars blasts out across their stellar neighbourhood, it drives the nebula's gas before it, creating shockwaves and helping to clear our a huge cavity-like region – a feature seen in many star-forming nebulae.

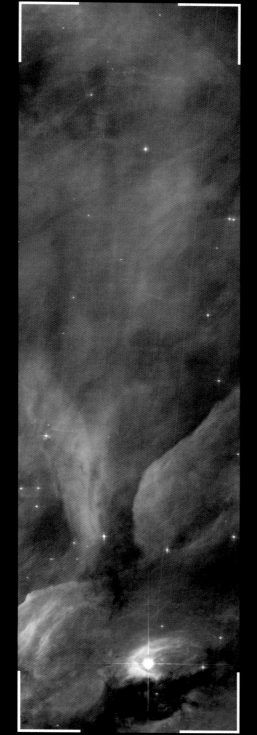

REFLECTION AND EMISSION

Not all of the light in glowing nebulae is a result of excitation and fluorescence – in some cases it is due to dust in the nebula simply reflecting starlight. This is the case with the bright blue-white glow around LP Orionis (at bottom left in the image above). Reflection nebulae typically look blue to telescopes on Earth because the light is reflected by tiny dust particles according to its wavelength. Only blue light, which has a shorter wavelength is 'scattered' through sharp enough angles to redirect it towards Earth.

<> 1,340 light years

>< 6 light years

CONE NEBULA

Lying some 2,600 light years from Earth in the constellation of Monoceros, the Cone Nebula is a dark mountain of star-forming gas and dust that forms just a small part of a larger nebula illuminated by S Monocerotis, the brightest and most massive star in the star cluster NGC 2264 that lies just off the top of these images. A torrent of radiation from this cluster is gradually wearing down the top of the cone, causing the stars that have recently formed within to emerge directly into view.

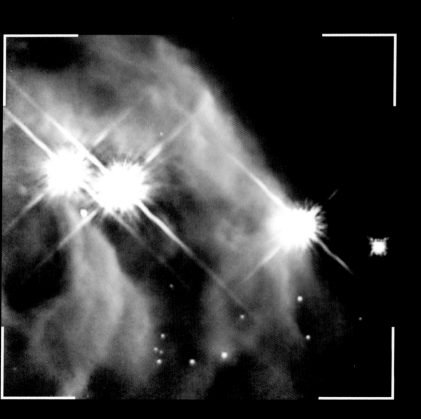

■ EMBEDDED STARS

A close-up image from Hubble's infrared NICMOS camera looks straight through the outer layers of warmer material to reveal the cooler regions within, showing a wealth of fine detail among the dust clouds. The brightest stars in the image are actually lying in the foreground, but the fainter yellow ones that appear at upper right and are only visible in the infrared may be either background stars or infant suns that are still forming within the Cone itself.

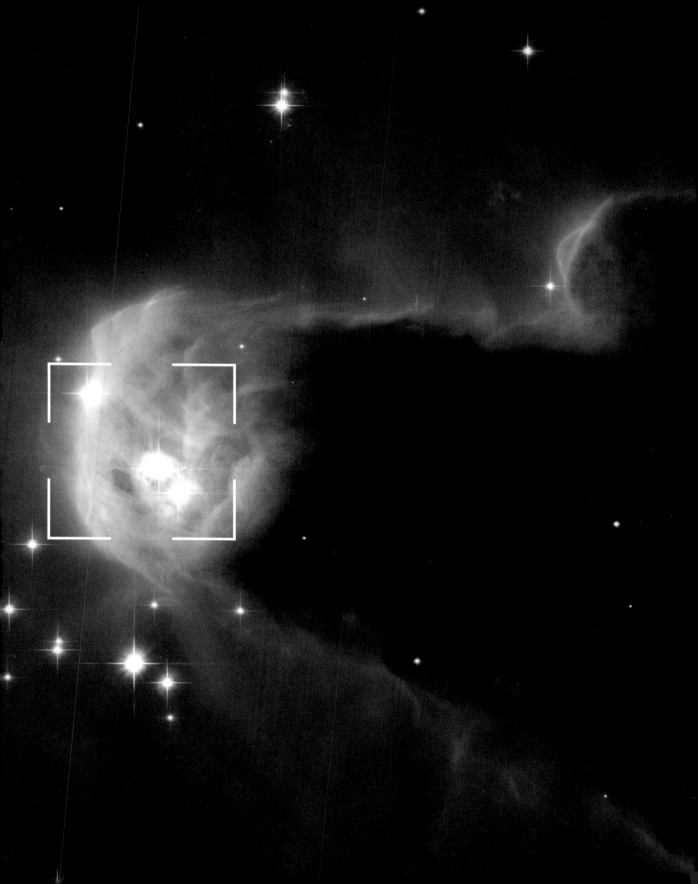

This delicate tower of star-forming material, known as the Spire, is being rapidly eroded by the radiation from young stars lying nearby. The result is a tendril of gas some 9.5 light years long. Stars are still forming within the Spire, but as the raw material or their formation is driven away, their growth will be stunted, and they will probably grow a little larger than our Sun – in this way, the first generation of stars born in any nebula chokes off the growth of their siblings.

THE SPIRE

Tenuous gas around one end of the Spire glows eerily thanks to excitation from the ultraviolet light of nearby stars. The Spire, it seems, contains denser clouds of hydrogen than its immediate surroundings, enabling it to withstand the erosive effects of radiation for longer. Shockwaves from the torrent of radiation sculpt the top of the spire into strange shapes, and may even trigger the collapse of knots of gas to form further stars.

69

EAGLE NEBULA 1

The Eagle Nebula is a huge cloud of nebulosity, catalogued as IC 4703, surrounding the newborn star cluster Messier 16, in the constellation of Serpens. Its general shape is that of an enormous star-filled cavity of glowing gas and dust. Radiation from the hot young stars of M16 excites gas within the cavity and causes it to glow, while denser, dust-rich regions form dark tendrils and pillars, including the eagle-like silhouette at its heart, and the famous 'Pillars of Creation'.

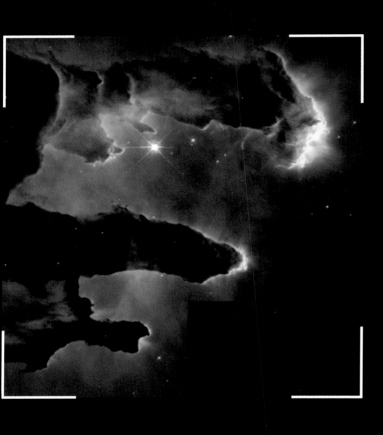

■ PILLARS OF CREATION

First photographed by the Hubble Space Telescope in 1995, the 'Pillars of Creation' form one of the most famous images produced by NASA's orbiting observatory.

The pillars gave astronomers their first really detailed view of the processes involved in star formation. Towering up to four light years high, they're believed to mark regions in which dozens of stars are forming from knots of gas that have grown dense enough to develop their own substantial gravity, and begin to pull in further material from their surroundings. As the growth of these young protostars snowballs, they will eventually grow hot and dense enough to shine.

<< 7,000 light years
>< 2 light years

■ EMERGING STARS

Although the EGG structures at first
appear to be bursting out of the pillar top
in this Hubble Space Telescope view, the
reality is rather different – the top of the
pillar is being eroded by fierce radiation
from nearby stars above it, and only the
more dense protostars can withstand the
erosion. As the bank of gas is worn down,
they are left intact. Each EGG casts a
protective shadow that blocks the radiation
from reaching the gas 'behind' it, allowing
it to stay connected tenuously to the main
body of gas, at least for a brief period.

Lying roughly 5,000 light years from Earth in the constellation of Sagittarius, the Lagoon Nebula is one of the most prominent star-forming nebulae in Earth's skies. Catalogued as Messier 8 or M8, it is a region of glowing gas roughly 100 light years across, with a young star cluster embedded at its heart. The spectacular image opposite, taken from the European Southern Observatory, covers an area of the sky with roughly eight times the diameter of the Full Moon.

■ TWISTERS

Although the gases within a star-forming nebula are far more tenuous than those in Earth's atmosphere, similar processes can sometimes operate. Deep within the Lagoon Nebula, the Hubble Space Telescope discovered these remarkable 'twisters' – spirals of gas and dust that are set spinning by rising currents of warm gas and strong hot 'stellar winds' in a very similar way to tornadoes on Earth.

5,000 light years
7 light years

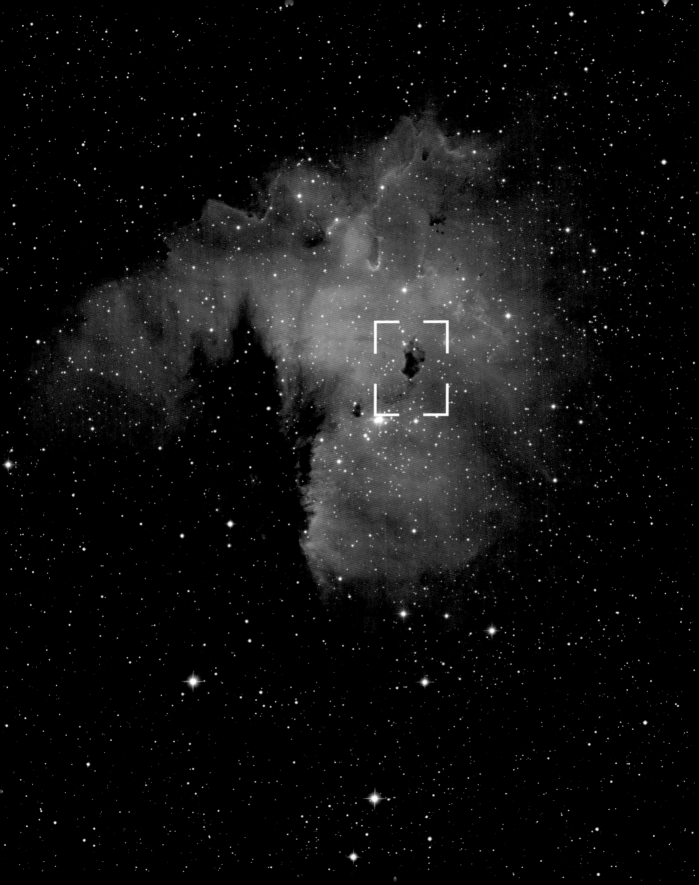

BOK GLOBULES

As clumps of star-forming material separate from larger pillars and mountains, they form small dark clumps, named Bok globules after Bart Bok, the astronomer who first discovered them and theorized that they were 'stellar cocoons' in the 1940s. Each globule typically contains either a single condensing core or a number of separate knots of gas that will typically form a binary or multiple star system and remain in orbit around one another throughout their lives. Bok globules are hard to spot except when they lie in front of a brighter background.

■ 'PAC-MAN' NEBULA

A single Bok globule, roughly two light years across, is silhouetted against the background glow of NGC 281, the so-called 'Pac-Man' nebula in the constellation Cassiopeia. As the centre of this cloud grows increasingly hot and dense, conditions will eventually become extreme enough to ignite the process of nuclear fusion, and a star will be born. The earliest stages of a star's life can only ever be revealed by infrared radiation, but as the star grows denser still and gains strength, it will eventually blow away the outer remnants of the Bok globule and emerge into full view.

PROPLYDS

As stars finally emerge from their cocoon-like Bok globules, they still carry a dense disc of gas and dust around their equator. These discs, known as proplyds or protoplanetary discs, arise as a natural consequence of the way that collisions between slowly rotating masses of gas tend to flatten out and develop structure. A detailed Hubble survey of the Orion Nebula discovered at least 42 newborn stars surrounded by such discs. Those closest to bright nearby stars heat up until they glow, while those further away appear dark, silhouetted against the nebula's brighter background gases.

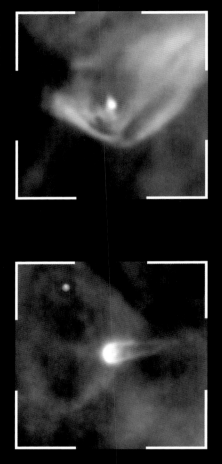

1,340 light years
0.02 light years

DISCS OF ORION

Most of the proplyds so far observed range up to 1,000 astronomical units in diameter (roughly ten times the diameter of our solar system's Kuiper Belt). However, they contain raw material that will coalesce into planetary systems over periods of hundreds of thousands, perhaps millions of years. Dust in the clouds is 'sticky' and clumps together, eventually coalescing into larger objects that develop their own gravity and pull in more matter from their surroundings, growing into 'protoplanets'.

[75] STELLAR JETS

As material from the surrounding gas cloud is pulled inwards onto a newborn star, it begins to spin more and more rapidly thanks to the concentration of mass – just as a pirouetting ice skater spins more rapidly as she pulls her arms in. Eventually, the star may spin so rapidly that it cannot hold onto all of its gas – instead, it sheds its excess material in tight jets that emerge from both poles, a so-called 'bipolar outflow'. As these jets shoot out across space, they may collide with neighbouring gas clouds, causing them to glow and creating a double-lobed nebula called a Herbig-Haro object.

■ HH-47
Embedded within a dark Bok globule in the constellation of Vela (the Sails), Herbig–Haro object HH-47 only becomes visible when viewed in infrared (heat) radiation. The image opposite shows a view from the Spitzer Space Telescope, while this close-up from the Hubble Space Telescope reveals 'wobbles' in the jets – perhaps caused as the central star is pulled in various directions by a large planet or an unseen stellar companion.

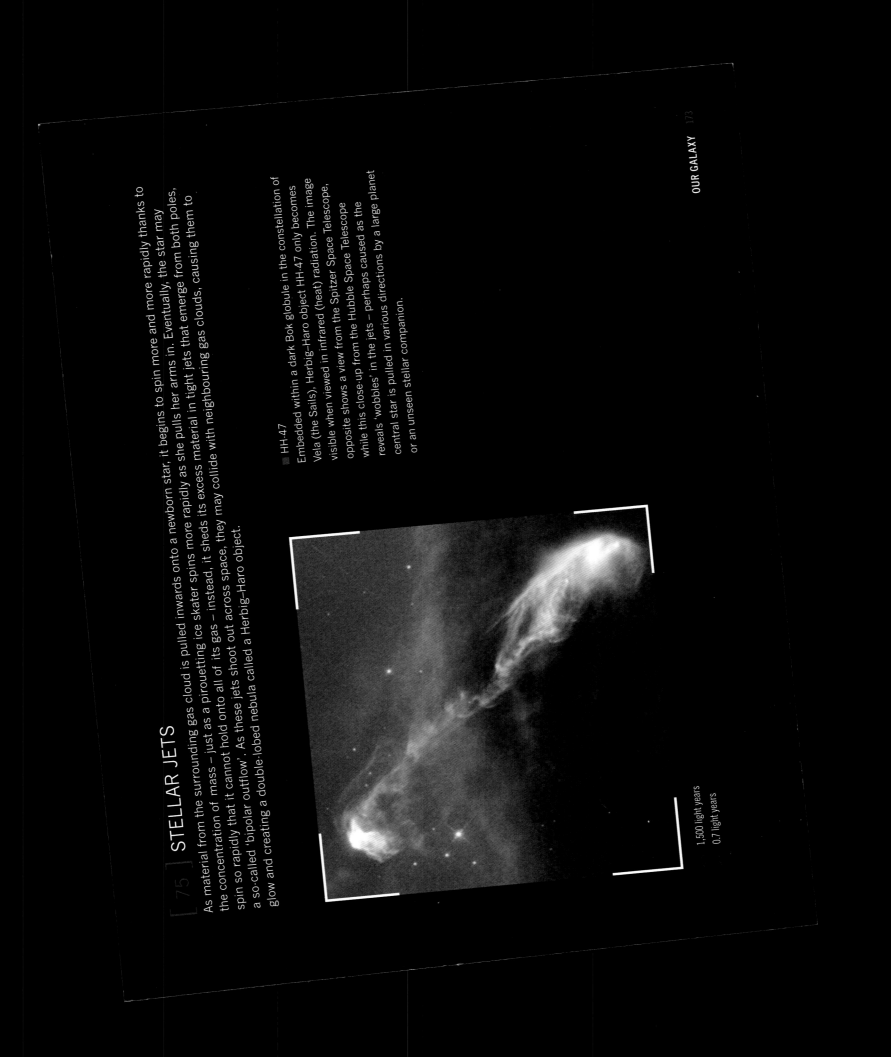

1,500 light years

0.7 light years

INFANT STARS

The long, slow process of star formation finally comes to an end once conditions in the star's core grow hot and dense enough for nuclear fusion to begin. As the nuclei of hydrogen atoms are forced together to create the next lightest element, helium, they release energy that helps support the star from within and brings its collapse to a halt. Particles streaming off the surface of the star create a fierce stellar wind that blows out into the surrounding space, as revealed by the young star LL Orionis in the Orion Nebula.

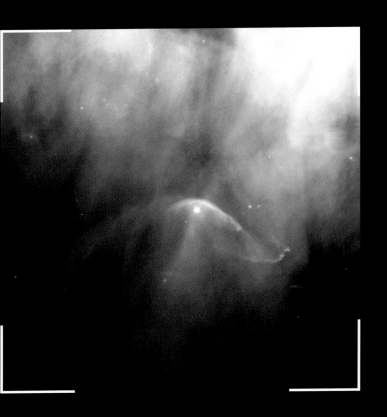

▨ STELLAR SHOCKWAVE

As LL Orionis drifts through the Orion Nebula, its strong stellar winds meet the slower-moving winds from other parts of the nebula head-on. The result is a glowing 'bow shock', similar to the wave produced in front of a fast-moving ship. In this case, the shockwave, illuminated by the energy from colliding particles, is at least half a light year wide. All stars are surrounded by a 'heliosheath' of stellar wind, but it's only in the rare case of stars like LL Orionis that it becomes visible.

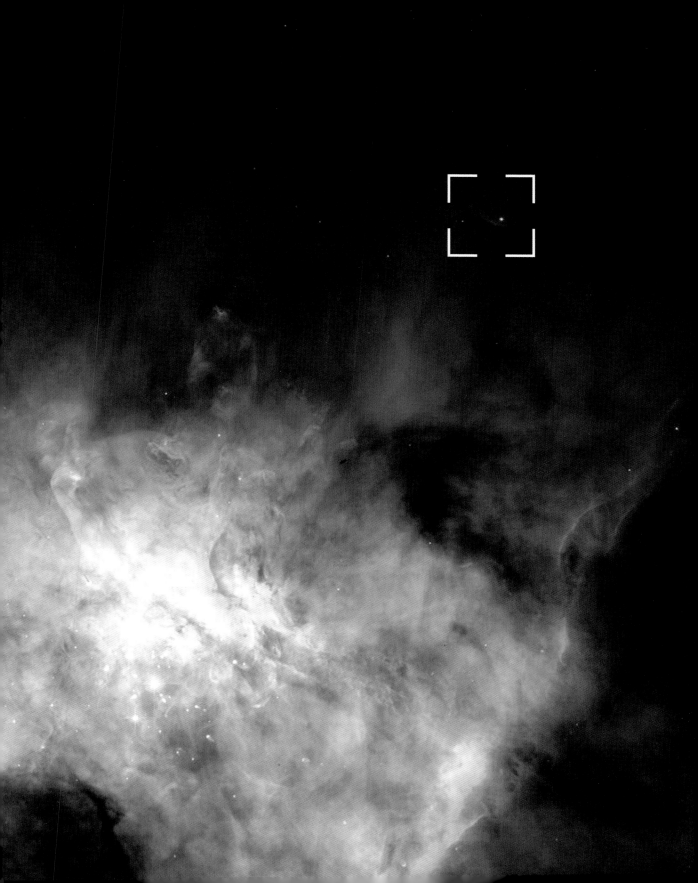

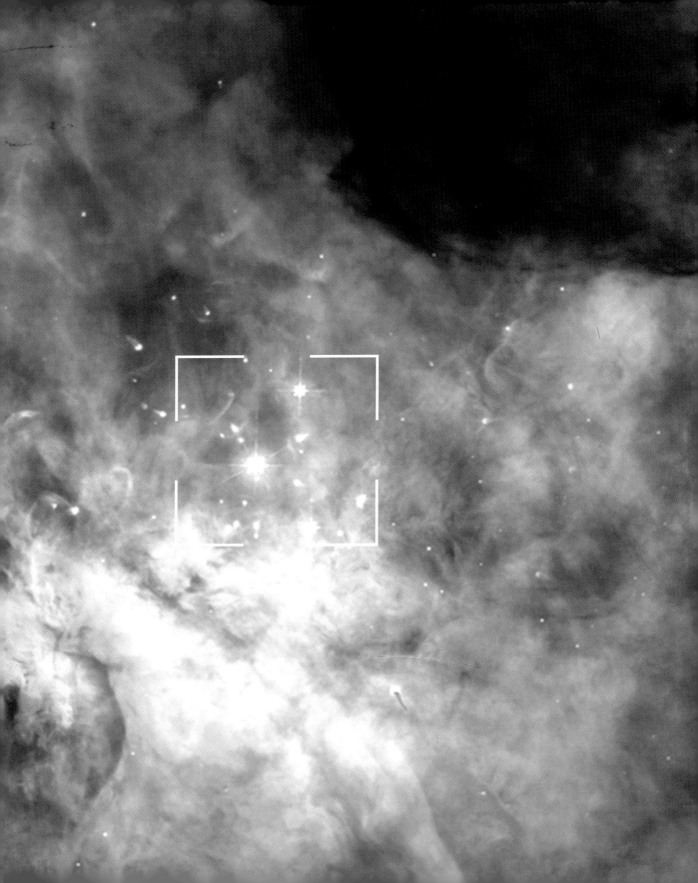

THE TRAPEZIUM

Star clusters, ranging in size from several dozen to hundreds of members, are an inevitable consequence of the way that stars form out of large gas clouds. Known as 'open clusters', these relatively loose associations of stars are doomed to gradually drift apart over hundreds of millions of years. The cluster at the heart of the Orion Nebula, dominated by a tight group of stars known as the Trapezium, is one of the youngest in the sky, with its stars still in the process of formation.

● INSIDE VIEW

The name 'Trapezium' naturally suggests a cluster of four stars, and this was indeed what 17th-century astronomers saw when they named it. However, we now know there are several more stars in the Trapezium, including some 'spectroscopic binaries' – stellar pairs that appear as a single star through even the most powerful telescopes, as seen in this Hubble image. These brilliant stars each weigh around 15–30 times as much as the Sun, and are doomed to live short but spectacular lives.

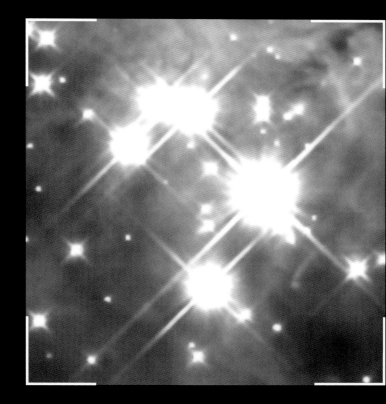

THE PLEIADES

Forming a fuzzy star cloud in the constellation of Taurus (the Bull), the Pleiades are probably the most impressive of all the sky's open clusters. The group is named in commemoration of the mythical 'Seven Sisters' of Greek mythology, and naked-eye observers can typically spot anything from six stars to a dozen or more. However, in total the Pleiades contain around a thousand stars scattered over a region of space 80 light years across. Over large distances, the cluster appears dominated by its largest members – heavyweight blue-white stars with ten times the mass of the Sun or more.

◼ MEROPE NEBULA

Long-exposure photographs reveal that the Pleiades are surrounded by the wisps of a blue 'reflection nebula', caused by dust scattering light from the cluster's brightest stars. The nebulosity is at its brightest around the star Merope. Despite appearances, the clouds surrounding the Pleiades are not remnants of the nebula in which they formed – instead, they are simply a patch of interstellar dust through which the cluster happens to be passing as it drifts away from the site of its birth.

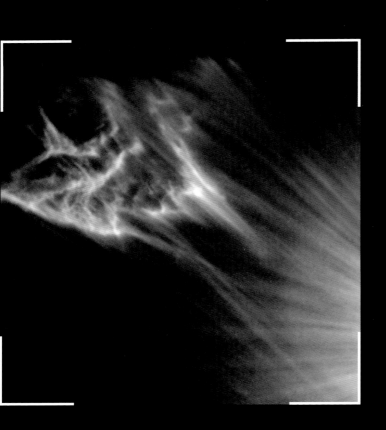

In contrast to the loose, short-lived groups of young stars that form open clusters, globular clusters are much larger, ball-shaped groups of older stars, containing many tens, even hundreds of thousands of low-mass, red and yellow stars. Globular clusters are often found close to the centre of galaxies, or in the 'halo region' above and below the disc of spiral galaxies – astronomers believe they are the building blocks of larger galaxies, and formed from the collapse of huge gas clouds early in cosmic history.

■ 47 TUCANAE

Lying at a distance of 16,500 light years from Earth, 47 Tucanae is one of the largest globular clusters in orbit around the Milky Way. Containing perhaps a million stars packed into a spherical ball 120 light years in diameter, to the naked eye it looks like a fuzzy star of moderate brightness, and even its name indicates that it was once catalogued as a star. The core of 47 Tucanae is one of the most tightly packed of all globular clusters, and Hubble observations of its core have shown that heavier stars tend to 'sink' towards the centre.

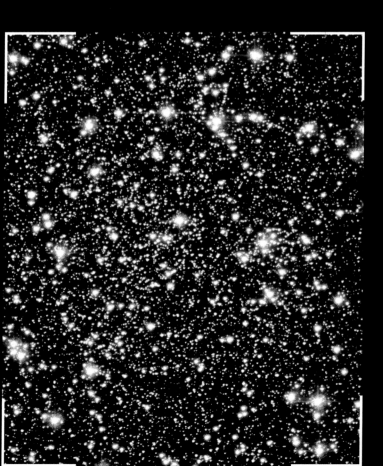

With an estimated 5–10 million stars and a diameter of 170 light years, Omega Centauri is by far the largest globular cluster in orbit around the Milky Way. On average, its stars are separated from one another by just 0.1 light years. In fact, Omega Centauri is so different from most other globular clusters that it may have originated in a very different way from its neighbours – astronomers suspect that it is the surviving core of an independent galaxy that was long ago pulled into orbit around the Milky Way and had its outer layers stripped away.

AGEING POPULATION

A Hubble view of Omega Centauri's core reveals an array of colourful stars, all between 10 and 12 billion years old. The majority are sedate yellow-white stars, not dissimilar to our Sun, but there are also red giants entering the latter stages of their evolution and faint, burnt-out white dwarfs. The brighter blue stars are a puzzle – any stars of this type should have long ago exhausted their fuel and disappeared in supernova explosions. Astronomers believe that these 'blue stragglers' may be the result of lower-mass, long-lived stars merging together in the crowded heart of the cluster.

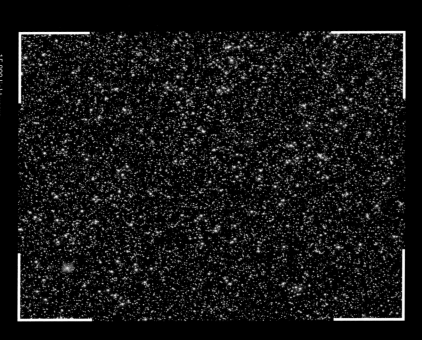

15,800 light years
6.3 light years

MULTIPLE STARS

While open clusters typically drift apart over tens of millions of years, some stars are born in close-bound pairs or groups, locked permanently in orbit around each other as 'binary' and 'multiple' systems. The star cluster Pismis 24, emerging from the nebula NGC 6357 in the constellation Scorpius, contains several examples of such stellar groups, most of which are so closely bound to one another that, from a distance of 8,150 light years, they appear at first to be single stars.

A HEAVYWEIGHT DOUBLE

Pismis 24-1 is the brightest member of this particular star cluster, and one of the heaviest multiple stars known. It consists of at least three stars, and this Hubble Space Telescope image shows the system split into two bright components (one of which is itself a 'spectroscopic' binary – a pairing that is only revealed by careful analysis of its spectrum). Each of these major stars is a 'blue supergiant' – a stellar heavyweight with 100 times the mass of the Sun and a surface temperature of around 50,000°C (90,000°F).

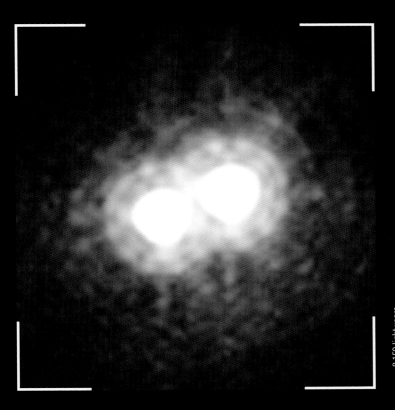

8,150 light years
0.1 light years

FOMALHAUT B

One of the brightest stars in southern skies, Fomalhaut, in the constellation Piscis Austrinus (the Southern Fish), appears to the naked eye as a normal white star, lying some 25 light years from Earth. However, astronomers have discovered that it emits unexpectedly large amounts of radiation in the infrared – a typical sign of a young star that is still surrounded by a disc of cool, dusty material left over from its formation. By blocking out the light of the central star itself, the Hubble Space Telescope was able to photograph this debris disc in stunning detail.

AN ALIEN WORLD

Astronomers believe that planets are born from the coalescence and collapse of discs, such as the one around Fomalhaut, into a small number of larger bodies. The sharp inner boundary at a distance of roughly 130 AU (about four times the diameter of Neptune's orbit) suggests this may already have happened close to Fomalhaut, and some astronomers have likened the disc to our own solar system's Kuiper Belt. Comparing Hubble images of the star from 2004 and 2006, astronomers spotted a planet of roughly Jupiter's mass following an 872-year orbit just inside the disc – the first such planet to be imaged in visible light. The image shown here reveals the motion of 'Fomalhaut b' over two years.

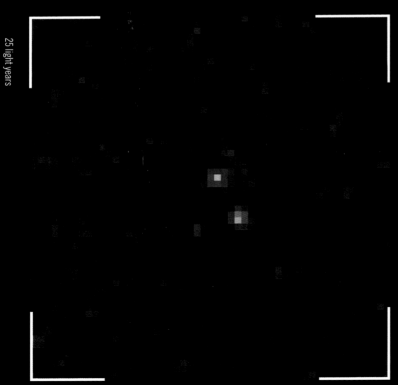

25 light years
1.5 billion km (930 million miles)

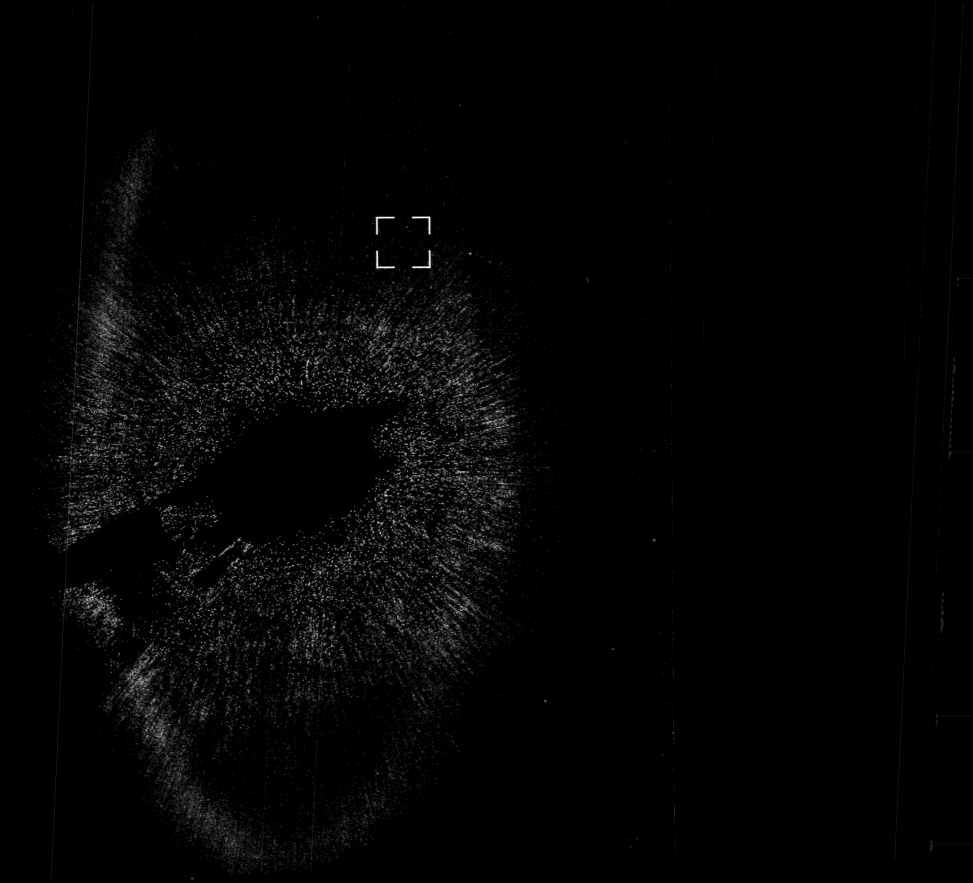

2M 1207B

Since the 1990s, astronomers have begun to discover planets around other stars in large numbers. Most only reveal themselves through minute changes to the spectrum of starlight, created as their parent star 'wobbles' back and forth under the influence of one or more planets. After some false alarms, the first confirmed image of an 'extrasolar planet' was made in infrared light from the European Southern Observatory's Very Large Telescope. '2M 1207b', as it is officially known, orbits an extremely feeble brown dwarf star called 2M 1207, 170 light years away in the constellation Centaurus.

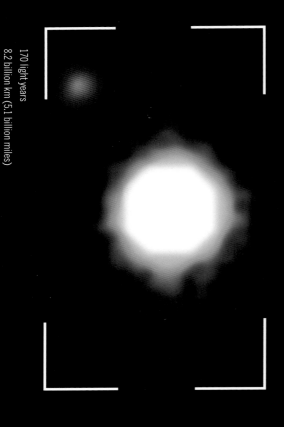

170 light years
8.2 billion km (5.1 billion miles)

RED-HOT PLANET

An artist's impression shows 2M 1207b as it might appear in close-up, orbiting through the dusty 'circumstellar disc' that also surrounds its parent star. The planet is a gas giant like Jupiter, but with between three and ten times the mass (the star 2M 1207, meanwhile, weighs about as much as 25 Jupiters – just one-fortieth of the Sun's mass). Infrared measurements show that 2M 1207b has a surface temperature of around 1,600°C (2,900°F). Since it orbits at about the same distance as Pluto from its feeble Sun, this cannot be a result of internal heating, so instead the planet must have an internal 'power plant' fuelled by gravitational contraction.

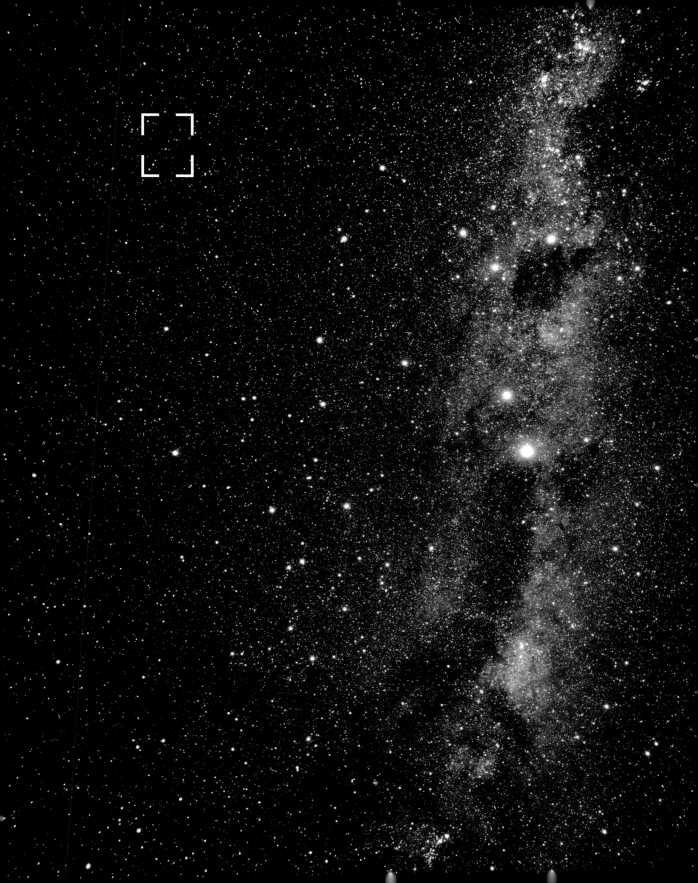

RED GIANTS

When a star begins to exhaust the fuel that has powered nuclear reactions in its core through most of its life, it undergoes physical changes that cause it to brighten considerably, while at the same time swelling enormously in size. As a result, its surface cools down and typically becomes orange or red. Because of their brilliance and colour, red giants are easily seen across great distances, and several are among the best known stars in the sky – for example, Betelgeuse, on the shoulder of Orion, is the ninth brightest star in the sky despite a distance of more than 600 light years from Earth.

BETELGEUSE

Red giants are among the largest stars known – even a relative stellar lightweight such as our Sun will swell to around the size of Earth's orbit in its red giant phase. Betelgeuse, which weighs 18 times more than the Sun, is a 'supergiant' roughly five times larger, and offers astronomers a rare chance to 'resolve' the surface of a star into a disc rather than a mere point of light. Taken in ultraviolet light, this Hubble image reveals a bright spot thought to mark one of the star's poles, and hints at Betelgeuse's 'fuzzy' outer envelope of gas.

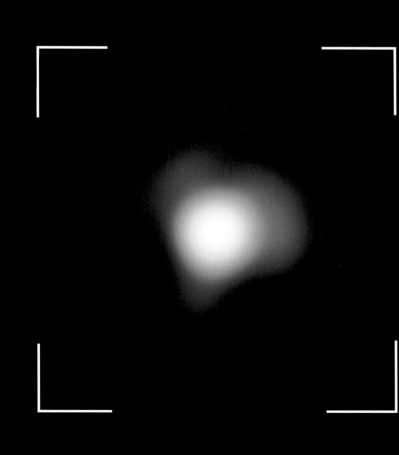

ETA CARINAE

While the Carina Nebula is best known as a site of abundant starbirth (see pages 149–153), it is also home to massive young stars that are already hurtling to their doom. At the centre of a region called the Homunculus Nebula lies a spectacular and unstable star system called Eta Carinae. Lying around 7,500 light years from Earth, the star itself is currently on the edge of naked-eye visibility, but it is unpredictable – in the 1820s and 1830s it brightened enormously, until in 1843 it was briefly the second brightest star in the entire sky.

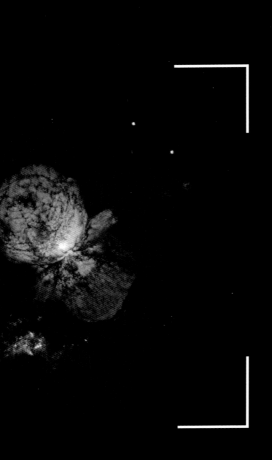

UNSTABLE BINARY

At the heart of the twin-lobed Homunculus Nebula, Eta Carinae itself was thought until recently to be a single star with more than 100 times the mass of the Sun, becoming increasingly unstable as it exhausted its fuel supplies and headed for a supernova explosion. Today it's thought to contain a pair of stars, but both are still massive in their own right, and their combined energy output is around 4 million times that of the Sun.

MIRA

As stars age and their internal structure changes, they frequently become unstable, changing their size and brightness and appearing in Earth's skies as 'variable' stars. One of the first such stars to be recognized is Mira (its name means 'wonderful'), in the constellation of Cetus (the Whale). Mira is a red giant that changes its brightness by a factor of up to a thousand in a period of 332 days. As it moves through space, its pulsations shed a trail of gas behind it, as revealed in the ultraviolet image opposite.

■ DISTORTED GIANT

Astronomers had long suspected that Mira was in fact a binary system, but it was only in 1997 that the Hubble Space Telescope successfully separated its two components. The variable red giant, Mira A, shown here in an ultraviolet image, has a companion that is faint but hot, and is almost certainly a white dwarf (see page 211). Close-up images of Mira A have revealed that the star is distorted by the gravitational pull of its companion, with a spiral of gas being pulled away towards Mira B.

> 420 light years

> 460–560 million km (285–350 million miles)

While some stars vary in regular periods and only change over periods of thousands or even millions of years, others undergo sudden, unexpected outbursts, typically known as novae and supernovae. One of the most intriguing and beautiful eruptions of recent times was the outburst of V838 Monocerotis, a previously unknown star in the constellation of Monoceros (the Unicorn), which exploded into life in 2002, briefly becoming a million times brighter than the Sun. Astronomers are still not sure what caused the eruption – it may have been a stage in the late evolution of a massive star, a merger between two smaller stars or even a star swallowing a giant planet.

■ TUNNEL OF LIGHT

One of the most impressive aspects of the V838 Monocerotis explosion was the eerie series of 'light echoes' it created over the next few years. These echoes, analogous to sound echoes in a cave, are caused as light from the explosion illuminates gas and dust in nearby space, before being reflected on to Earth. Because light travels at a limited speed, reflected light takes longer to reach Earth than that coming direct from the explosion. So as the outburst itself faded away (captured in the initial Hubble image shown at left), astronomers were treated to a series of beautiful reflections off the surrounding 'interstellar medium'.

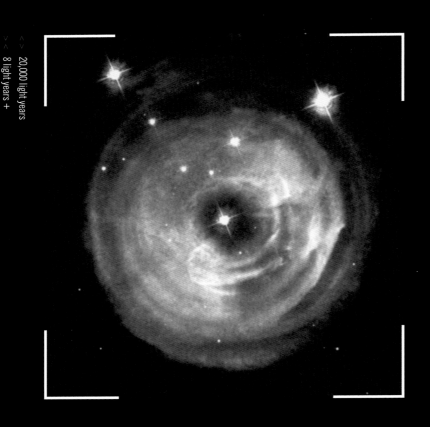

< > 20,000 light years

> < 8 light years +

Stars of roughly solar mass go through two distinct red giant phases as they near the end of their lives and burn through their supplies of first hydrogen, and then helium. At the end of the second giant phase, a Sun-like star swells to enormous size and becomes unstable, expanding and contracting, and eventually puffing off its outer layers in a series of interstellar 'smoke rings'. The rings continue to glow for some time, illuminated from within by the remnants of the central star, which grows increasingly hot as deeper and deeper layers are exposed. The result is a short-lived but beautiful 'planetary nebula'.

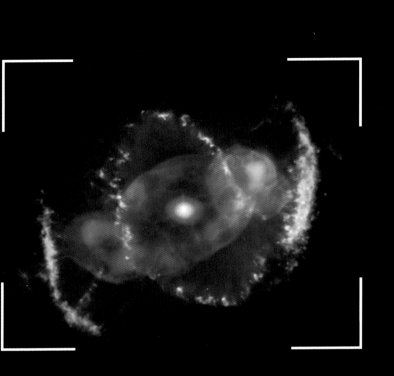

■ CAT'S EYE NEBULA

Depending on the exact conditions in which they form, planetary nebulae can develop in a range of elaborate shapes. The Cat's Eye Nebula, NGC 6543 in the constellation of Draco (the Dragon), is one of the most complex, and intensely studied, planetaries in the sky. It consists of a series of intersecting ellipses of gas, glowing in a range of different colours and emitting radiation from the cool infrared to high-energy X-rays. The X-rays revealed in this combined Hubble/Chandra X-ray Observatory image create the bluish glow in the Cat's Eye's elongated 'pupil' – they originate where a fast 'stellar wind' from the central star catches up and collides with slower-moving gas ejected in a previous phase.

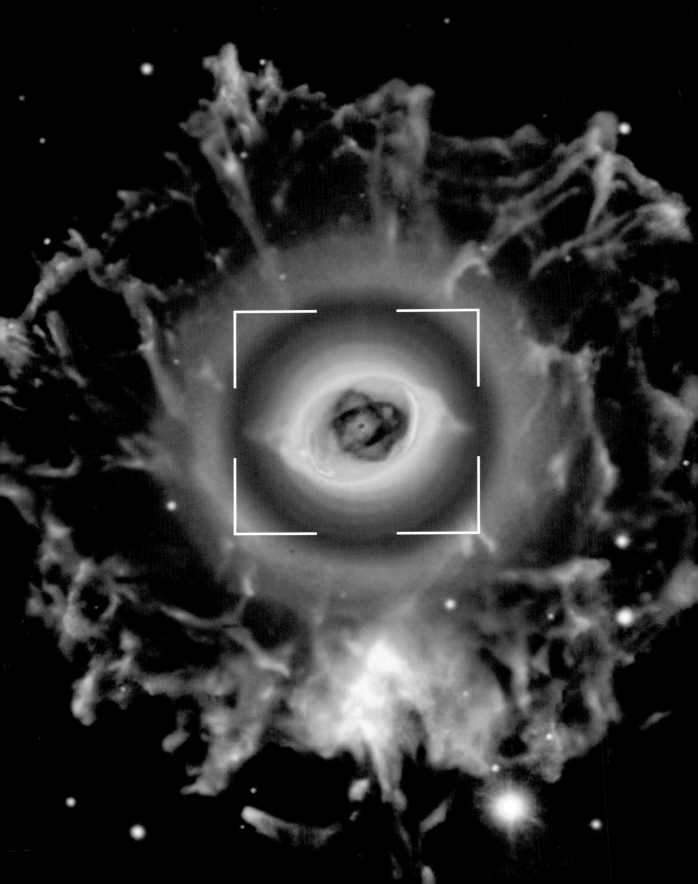

Computer-processed images from sensitive ground-based telescopes reveal that the Cat's Eye covers a huge volume of space. The outer layers revealed in the false-colour image opposite are rich in oxygen (green and blue) and nitrogen (red). Close to the centre is a system of more orderly concentric rings, created by a series of 'shells' that formed during regular pulsations of the central star that occurred between 15,000 and 1,000 years ago. The brighter central structure has developed in just the last few centuries.

■ COMPLEX STRUCTURE

This beautiful Hubble close-up shows the innermost regions of the Cat's Eye – the complex interlocking shape at the centre is caused by expanding, interlocking bubbles of gas inflated from within by the star's hot stellar winds. The bubbles are pinched together by a dense, more slowly expanding disc around the 'waist'. The central axis of the system seems to have changed over time, and this is good evidence that the central star has an unseen companion that causes it to 'wobble' over time.

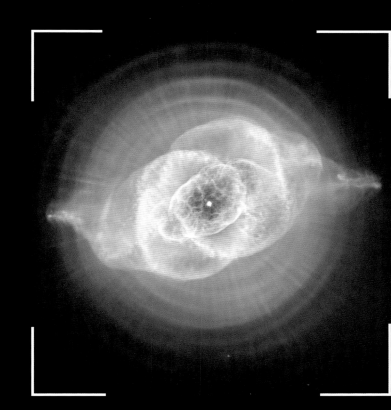

Located around 2,000 light years from Earth in the northern constellation of Lyra (the Lyre), the Ring Nebula is one of the brightest and most easily seen planetary nebulae. It is also one of the most elegant, appearing in visible light as a slightly elliptical 'smoke ring'. Astronomers used to think that the ring was a bubble that only became visible at its edges, but it now seems to have a cylindrical structure that just happens to point towards Earth. Infrared observations such as the image shown opposite have recently revealed complex loops of cool gas extending beyond the visible ring.

■ LAKE OF LIGHT
The Ring Nebula looks like an inviting tropical pool in this Hubble Space Telescope image. Different gases within the nebula, excited by ultraviolet radiation from the central star, are responsible for different colours of light, giving rise to the Ring's beautiful multi-hued appearance – the blue haze in the centre comes from helium, while oxygen and nitrogen around the edges glow green and red respectively.

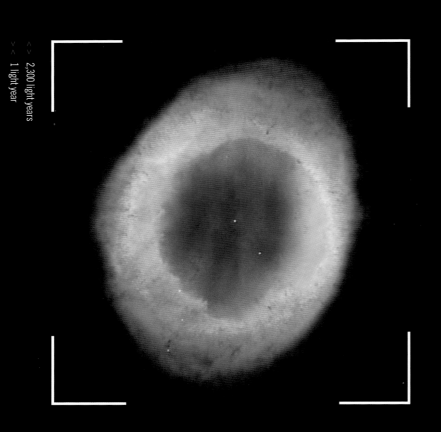

< > 2,300 light years
> < 1 light year

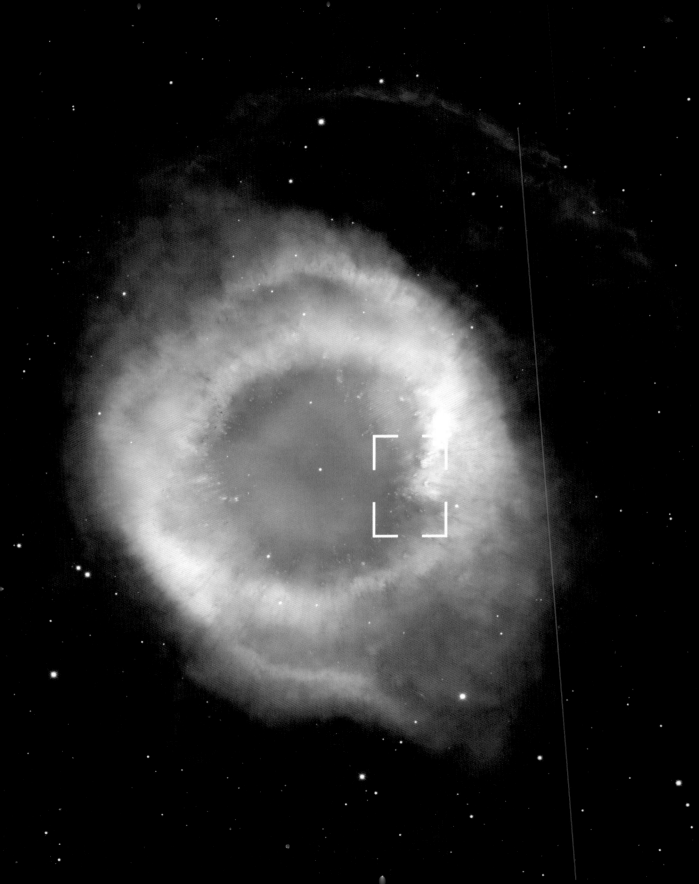

HELIX NEBULA 1

At a distance of around 700 light years from Earth, the beautiful Helix Nebula is one of the closest planetaries to Earth. With an apparent size in the sky close to that of the Full Moon, it is also extremely large, with a diameter of about five light years. Though difficult to observe with small telescopes because its light is 'smeared' across a large area, beautiful Hubble images have made it famous and given rise to the popular nickname, 'Eye of God'.

'COMET' KNOTS

Some of the most striking details in the Helix are knots of denser gas that appear to stream towards the centre. Nicknamed 'comets', these gaseous spokes actually have heads the size of our entire solar system. They are formed as a hot stellar wind from the central star ploughs into a dense outer shell of slower-moving gas, blowing through it and pulling it out into long radial streamers.

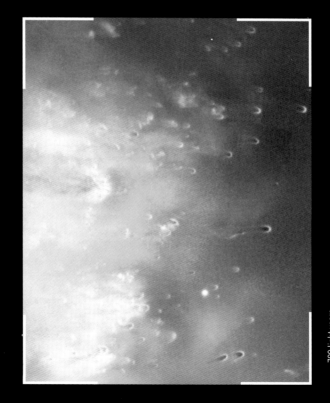

700 light years

0.75 light years

An infrared image of the Helix Nebula from NASA's Spitzer Space Telescope (opposite) reveals hidden features related to the temperature of matter within. The red regions radiate at comparatively cool wavelengths, while green comes from gas at intermediate temperatures and blue from the hottest areas. The nebula's central white dwarf star is the tiny white dot at the centre, surrounded by a red dust ring. The brightest, bluest regions of the Helix itself reveal heating effects from shockwaves within the outer reaches of the nebula.

700 light years

0.75 light years

■ SHOCK FRONT

According to estimates of the Helix's current size and the expansion rate of its outer regions (around 40 km/s or 25 miles per second), the nebula probably began to form around 12,000 years ago. Collisions between masses of gas moving at different speeds give rise to glowing shockwaves revealed in this Hubble image. Blue and green colours in the nebula's interior indicate the presence of oxygen, while red light in the outer layers originates from hydrogen and nitrogen.

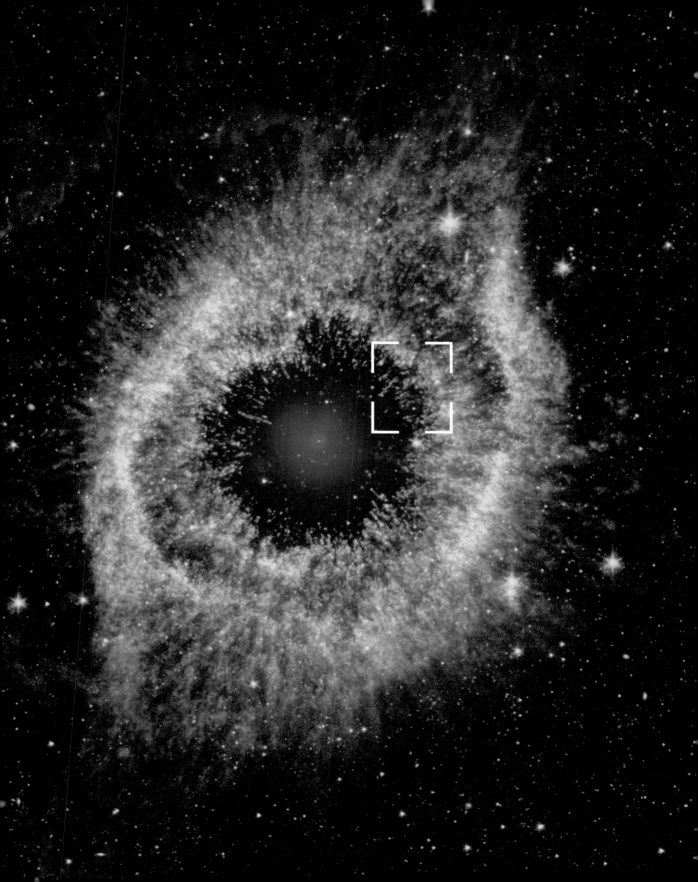

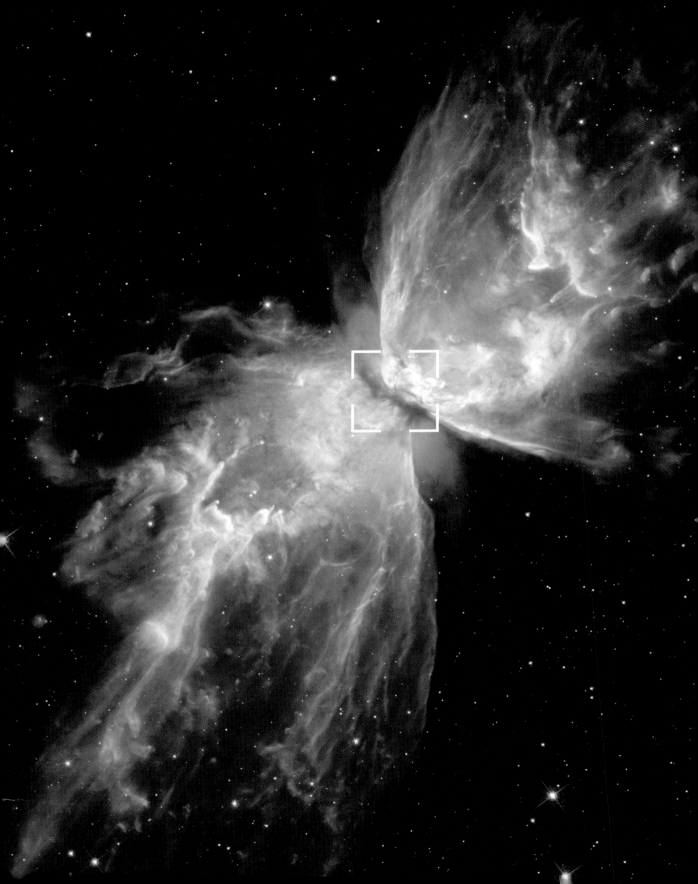

In contrast to the sedate-looking Ring and Helix Nebulae, the Bug Nebula in the constellation of Scorpius (the Scorpion), seems like a very different, and far more explosive, place. But the difference between the three objects comes largely from our point of view, and the Bug, too, is a planetary nebula. In this case, we are seeing the nebula sideways-on, so that expanding bubbles of gas on either side of the central star are clearly visible. Further out, there are traces of two fainter bubbles from a previous phase of expansion.

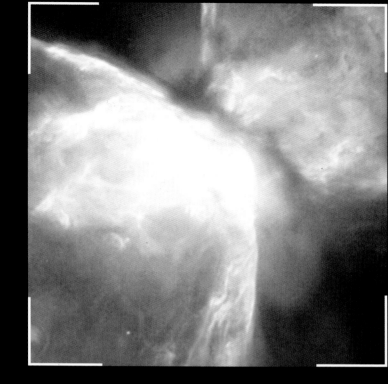

DARK DISC

The 'bipolar' appearance of the Bug Nebula is caused by a dense ring of dust and gas that lies almost edge-on to Earth at the centre of the nebula, constraining the expansion of gas away from the central star hidden within it. Studies of this dust have revealed that it contains a complex array of chemical compounds, including carbon-based molecules and icy 'hailstones'. Many such complex chemicals are thought to be formed in the upper atmospheres of red giants as they age, and are later expelled during the formation of planetary nebulae.

WHITE DWARFS

Once a Sun-like star has shed its outer layers, all that remains behind is a dense, hot, slowly cooling core. Roughly the size of the Earth, this 'white dwarf' star consists of compressed atomic nuclei – a teaspoon of its material weighs several tonnes, allowing white dwarfs to exert powerful gravity despite their size. Indeed, it was gravitational influence that allowed the first white dwarf, Sirius B, to be discovered – locked in a 50-year orbit around Sirius A, the brightest star in the sky, it creates telltale 'wobbles' in Sirius A's slow path across the sky.

∨∧ 8.6 light years
∧∨ 600 million km (390 million miles)

■ THE SIRIUS SYSTEM

Lying on our cosmic doorstep at a distance of 8.6 light years, Sirius B is nevertheless hard to observe thanks to the brilliance of Sirius A itself, as revealed in this Hubble Space Telescope image. The white dwarf (at lower right of the brighter star) weighs roughly the same as the Sun, and half as much as Sirius A, but is far smaller than a normal star. However, Sirius B's surface temperature of around 25,000°C (45,000°F) is more than twice that of Sirius A. Since stars evolve more rapidly the heavier they are, Sirius B must once have been far more massive than its neighbour, outshining it in Earth's skies before discarding most of its material in a planetary nebula that has long since faded and dispersed.

CRAB NEBULA 1

In 1054, a brilliant new star appeared for several months in Earth's skies, and was recorded by astronomers around the world. This stellar outburst was a supernova, the explosive demise of a star far more massive than the Sun, around 6,500 light years away. The site of that explosion, in the constellation of Taurus (the Bull), is today marked by a faint nebula first discovered in 1731 and later listed as 'Messier 1' in the catalogue of non-stellar objects compiled by French astronomer Charles Messier. Today, M1 is better known as the Crab Nebula, one of the sky's finest supernova remnants.

SHREDDED REMAINS

A supernova occurs when a star far more massive than the Sun finally exhausts the fuel supplies in its core and collapses violently under its own weight. As the core collapses, a shockwave rips through the outer layers and creates conditions in which they can erupt in a tremendous burst of nuclear fusion. The shredded remains of the star, still rapidly expanding today, consist of a wide variety of elements including carbon, nitrogen, oxygen, iron, neon and sulphur, and have temperatures of up to 18,000°C (32,500°F).

6,500 light years
3 light years

Hidden at the heart of the Crab Nebula lie the shrunken remains of the star that created the original supernova explosion, compressed into a superdense 'neutron star' (see page 218) that may contain the mass of several Suns within a city-sized volume. Thanks to their compact nature, neutron stars spin incredibly quickly and have powerful magnetic fields that channel most of their radiation into two narrow beams emerging from their magnetic poles. As the neutron star rotates, these lighthouse-like beams sweep across Earth every 33 milliseconds, creating a flashing cosmic beacon known as a pulsar.

■ PULSAR WINDS

Most astronomical objects only change their appearance over very long timescales, but the extreme nature of its central pulsar makes the Crab Nebula a notable exception. The images on these pages combine visible-light observations from the Hubble Space Telescope with X-ray images from the Chandra X-Ray Observatory, and record rapid changes in the 'ripples' surrounding the Crab Pulsar. The ripples are in fact shockwaves caused as a stream of superhot particles blowing off from the pulsar's equator and slamming into gas in the surrounding nebula.

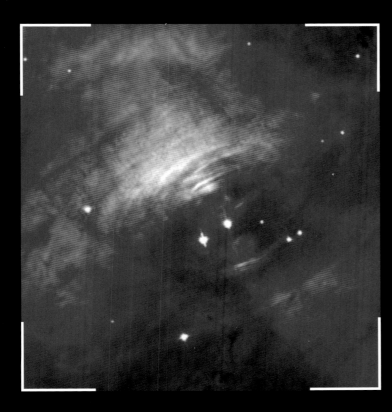

6,500 light years

3 light years

In contrast to the widely observed Crab Supernova of 1054, the most recent supernova explosion in our galaxy went strangely unrecorded. Its remnant is the most powerful compact radio source in the sky beyond our solar system, but it was only discovered and catalogued as 'Cassiopeia A' in 1947 during the early days of radio astronomy. The expanding supernova remnant, meanwhile was not observed in visible light until 1950. By measuring the rate of the remnant's expansion, astronomers have calculated that the supernova exploded around 1670 – interstellar dust clouds probably prevented its light reaching Earth at the time.

COMPLEX REMNANT

This false-colour composite combines the visible-light image from the Hubble Space Telescope (opposite, shown here in yellow), with infrared observations from the Spitzer Space Telescope (in red) and an X-ray view from the Chandra X-Ray Observatory in green and blue. The outer cloud of blue filaments emanates from multi-million-degree temperatures created as matter from the expanding supernova collides with the surrounding interstellar medium. Cassiopeia A's central neutron star is the cyan dot near the bottom of the picture.

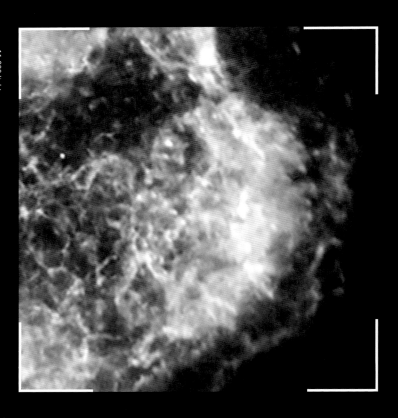

11,000 light years
6 light years

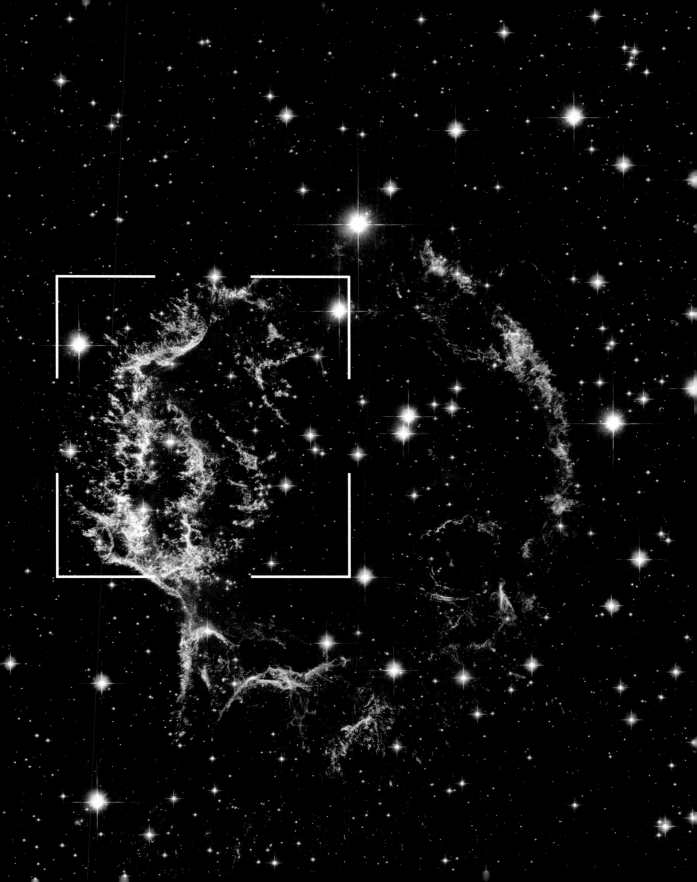

NEUTRON STARS

While the slow collapse of a Sun-like stellar core usually leaves the atomic nuclei intact within it, the far more violent collapse seen during a supernova breaks them apart, compressing the entire stellar core into a superdense soup of subatomic particles known as neutrons. So-called neutron stars are the densest objects in the Universe, typically containing between 1.4 and 5 solar masses of material compressed into a ball just a few kilometres across. With surface temperatures of more than 100,000°C (180,000°F), they emit powerful X-rays and other forms of radiation, which are often channelled into beams to form a pulsar (see page 215).

PSR B1509-58

The ghostly 'hand' nebula surrounding the neutron star and pulsar PSR B1509-58 is a cloud of high-energy particles imaged by the Chandra X-ray telescope. The pulsar itself lies at the base of the 'palm' – electrically charged particles streaming off its surface become tangled up in a magnetic field roughly 15 million million times more powerful than Earth's and accelerated to high speeds. They then emit excess energy in the form of X-rays.

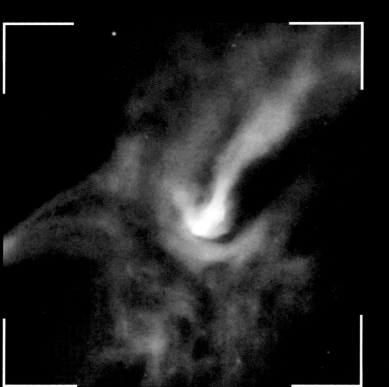

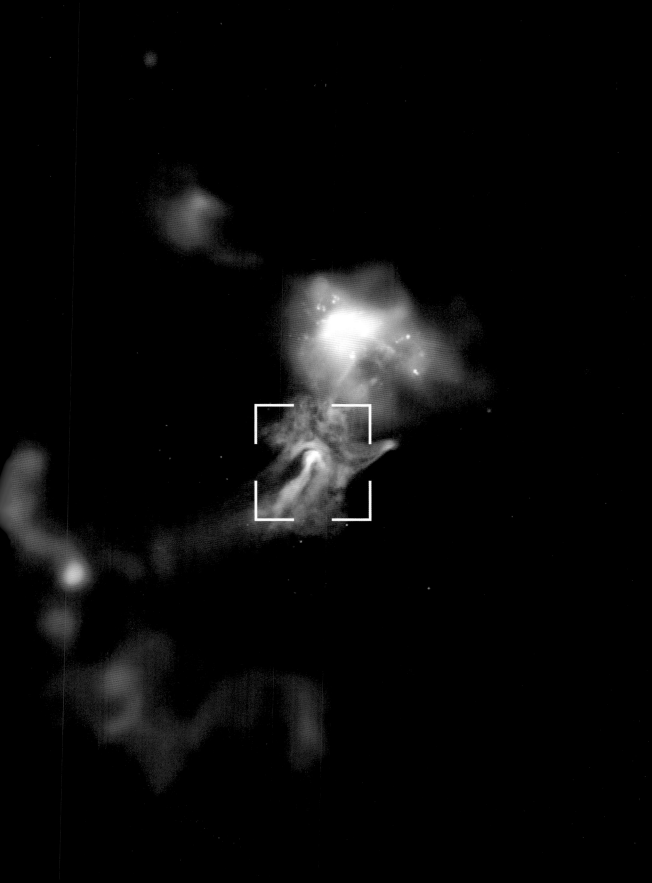

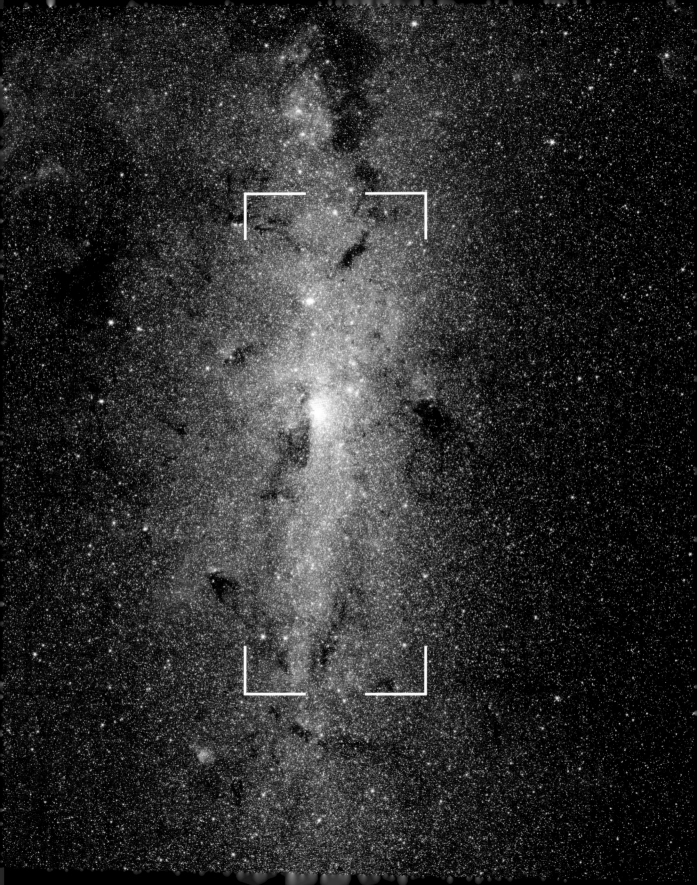

From our point of view on Earth, the star-crowded disc of our galaxy wraps its way around the sky as a pale band – the original Milky Way. The star clouds are densest in the direction of the constellation of Sagittarius, and the centre of our galaxy. Here a huge flattened bulge containing billions of old red and yellow stars, surrounds the remarkably crowded region of the 'galactic centre' itself. While the intervening stars make this region impenetrable in visible light, infrared images such as the one opposite reveal a wealth of detail in the form of dust clouds and hot ionized gas.

CHANDRA'S VIEW

An image from the Chandra X-ray Observatory reveals a very different side to the region around the galactic core. Colours from red through green to blue indicate X-rays of increasing energy, producing a multi-coloured panorama of various 'compact' X-ray sources, bathed in the white haze from huge clouds of multi-million-degree gas. The compact sources are associated with a variety of superdense stellar remnants, including white dwarfs, neutron stars and stellar-mass black holes (superdense points in space from which not even light can escape, formed during the death of the heaviest stars of all).

✧ 26,000 light years

✕ 700 light years

SAGITTARIUS A*

At the very centre of our galaxy lies a complex radio source known as Sagittarius A, with three separate elements – diffuse clouds known as Sagittarius A East and West, and a compact object called Sagittarius A*. This central object, around which everything else in the region ultimately orbits, is believed to be a supermassive black hole that packs the mass of 4 million Suns into a region with a diameter no larger than the orbit of Uranus. Although it is impossible to observe the black hole directly, the rapid orbits of nearby stars give away its presence, and astronomers now believe that similar objects lie at the centre of most, if not all, large galaxies.

ECHOES OF ACTIVITY

The supermassive black hole at our galaxy's centre exerts a powerful gravitational influence over everything around it, but at present most nearby objects maintain a safe distance and avoid being pulled to their doom. However, scattered particles falling into the black hole release energy as they go, explaining the existence of the Sagittarius A* radio source. But the black hole's behaviour is unpredictable, and this Chandra Observatory view reveals X-ray 'light echoes' off nearby gas clouds (see page 196), revealing that it burst briefly into life as it swallowed a Mercury-sized mass around 50 years ago.

26,000 light years
60 light years

BEYOND THE MILKY WAY

Despite their immense distance, the latest advances in Earth- and space-based telescope technology allow us to study them in unprecedented detail.

Galaxies are huge collections of stars, gas and dust, growing to enormous size and capable of producing many generations of stars. Ranging in size between dwarfs a few thousand light years in diameter and enormous spirals and elliptical balls that may be more than 100,000 light years across, they are nevertheless separated by comparatively small distances of hundreds of thousands or millions of light years. This means that, in comparison to smaller objects such as stars or planets, galaxies are quite closely packed together. As a result, they frequently interact through the far-reaching influence of their gravity, and are prone to direct collisions and mergers that change their structure and content over billions of years. They also have a tendency to form larger, gravitationally bound clusters and superclusters that are the largest 'objects' in the Universe.

Our own galaxy is a comparative heavyweight – a spiral system with a disc roughly 100,000 light years in diameter and an overall mass equivalent to 700 billion Suns. It is one of three spirals that dominate our own small galaxy cluster, the Local Group – the others are the Andromeda Galaxy (see pages 244–247), which is even larger but has around the same mass, and the far smaller Triangulum Galaxy, just half the diameter of the Milky Way. These three galaxies rule over a region of space some 10 million light years across, influencing the paths of several dozen smaller dwarf galaxies. Many of these dwarfs, including the substantial Large and Small Magellanic Clouds, are locked in orbit around one or other of the larger galaxies, but others are far more remote.

Astronomers recognize three major classes of galaxy existing in the Universe today, with several distinct types within each class. Spirals like the Milky Way account for roughly a quarter of nearby galaxies, and range from tightly wound, well-defined 'grand design' spirals, to far looser 'flocculent' ones. All spirals consist of a central bulge of old red and yellow stars, surrounded by a wide disc of stars, gas and dust. Spiral arms – regions of star formation that stand out because they are filled with bright star-forming nebulae and brilliant open star clusters – wind their way out from the central bulge to the edge of the disc. The spiral arms are not physical structures – if they were, they would 'wind up' and wrap themselves around the bulge in just a few rotations. Instead, the spiral regions are 'density waves' caused by the overlap of countless individual stellar orbits around the galactic centre. Material in the galaxy's disc follows more or less circular orbits, but slows down and becomes more densely packed as it enters the denser region, rather like cars entering a traffic jam. This compression is what triggers the waves of starbirth in spiral arms – while they appear permanent, in reality huge numbers of individual stars are moving in and out of them all the time. Density waves can give rise to many complex spiral patterns, and the external influence of other galaxies can complicate matters further, strengthening or distorting the waves or frequently producing a long straight 'bar' of stars that crosses the galaxy's bulge and has the spiral arms rooted at its ends.

Compared to the complex spiral galaxies, the other major forms of galaxy are relatively simple in

GALACTIC NEIGHBOURHOOD

This artist's impression shows the immediate vicinity of the Milky Way and the nearby galaxies of the Local Group. The Milky Way itself is the large spiral galaxy at lower centre, while the group's other major spirals, the Andromeda and Triangulum galaxies, lie at upper centre and upper left respectively. Surrounding each is a halo of smaller satellite galaxies – irregulars, small ellipticals and dwarf spheroidals – while a handful of more independent dwarf galaxies loiter around the edges. The Local Group is a system bound together by its own gravity, and the Milky Way and Andromeda are being drawn towards one another at a speed of around 120 km/s (75 miles per second). They will ultimately collide in around 5 billion years time.

0

10 million light years

structure. Ellipticals are huge balls of mostly old red and yellow stars with overlapping orbits that combine to create a structure that can range from perfectly spherical to an elongated cigar shape. They typically contain very little in the way of star-forming gas and dust, which explains why their surviving stars are the relatively long-lived ones with lower masses. Ellipticals comprise about 20 per cent of galaxies in the nearby Universe, but far more within dense galaxy clusters. They vary hugely in size and density, from small but sparse 'dwarf spheroidals' and more densely packed but still compact ellipticals like the ones orbiting the Andromeda Galaxy, through larger galaxies similar in mass to the Milky Way, up to truly enormous 'giant ellipticals' with the mass of up to a trillion Suns that are only found in the centre of large galaxy clusters.

Lenticulars, meanwhile, are a rare type of galaxy with properties somewhere in between those of ellipticals and spirals – they have elliptical-type cores surrounded by a disc of gas and faint average stars, but no spiral arms.

The final widespread form of galaxy are the irregulars – small-to-medium-sized galaxies that, as their name suggests, are comparatively structureless. In stark contrast to the ellipticals, irregulars are rich in the raw materials of star formation, with large numbers of bright young stars and brilliant open clusters. Some of the richest and brightest irregulars, undergoing huge waves of starbirth, are known as 'starburst' galaxies. Irregulars account for around a quarter of galaxies in the nearby Universe, but become the dominant type when we look at regions of space billions of

years away representing much earlier periods of cosmic history.

Ever since Edwin Hubble conclusively proved the existence of galaxies beyond our own, and drew up the first classification schemes in the 1920s, astronomers have suspected that the different galaxy types have an evolutionary link – in other words, they represent different stages in the life of a typical galaxy. However, early opinions on the 'life cycle' of galaxies have been turned on their heads, and it now seems that the true story is far more complex than previously imagined. The earliest galaxies were small irregulars that grew by absorbing cool gas from their surroundings, and ultimately coalesced to form more structured spirals. Drawn together in the heart of galaxy clusters, spirals merge together in turn to form large ellipticals. Collisions between different types of galaxy often give rise to 'peculiar' galaxies with structures that don't conform to any of the standard groups, and can also trigger violent activity around the central black hole that lies at the heart of most galaxies.

The merger process between spirals drives much of their star-forming gas away into intergalactic space and leaves the galaxy unable to form new stars. However, in some cases the galaxy can draw in enough gas from its surroundings to begin star formation again, passing through a lenticular phase on its way back to becoming a true spiral.

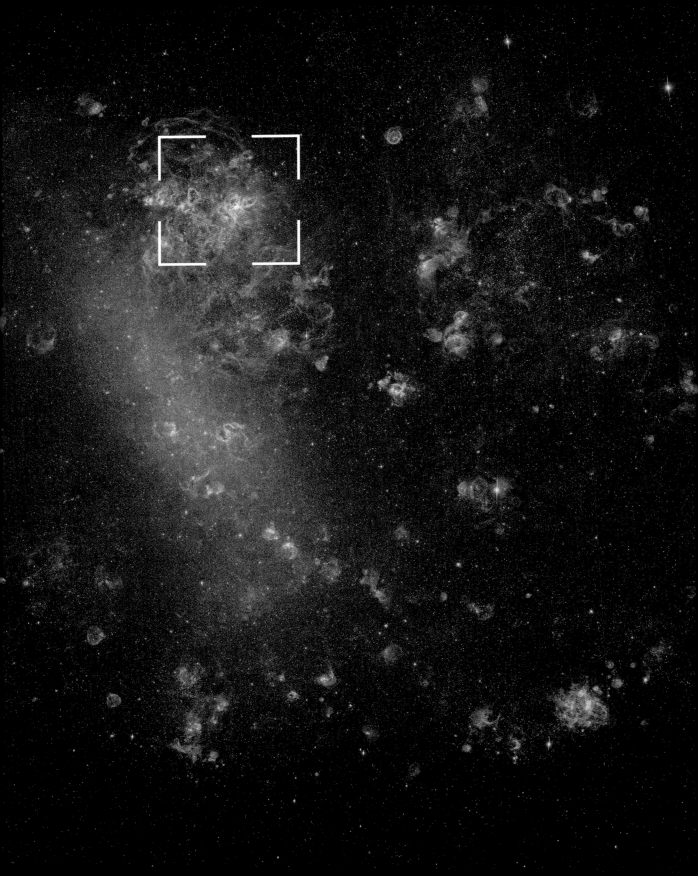

| LARGE MAGELLANIC CLOUD

The largest and brightest of a dozen or so small galaxies in orbit around the Milky Way, the Large Magellanic Cloud (LMC) is a prominent object in southern skies, appearing as a separate 'blob' of the band of the Milky Way in the constellation of Dorado (the Swordfish or Dolphinfish). Although it appears fairly amorphous at first glance, astronomers have noted some traces of structure, including a central 'bar' of stars and what appears to be a single spiral arm. It's possible that the LMC was a small spiral in its own right before being disrupted as it was pulled into orbit by our own galaxy.

■ TARANTULA NEBULA

The most prominent feature within the LMC is a spectacular region of star formation known as the Tarantula Nebula. This is the largest and brightest such nebula in the entire Local Group of galaxies, with a diameter of around 650 light years – if it was as close to Earth as the Orion Nebula (see pages 154–157), it would appear bright enough to cast shadows. The Tarantula may owe its size to a location on the LMC's 'leading edge' – as the galaxy ploughs through clouds of intergalactic matter, its gas is compressed to give birth to huge quantities of stars.

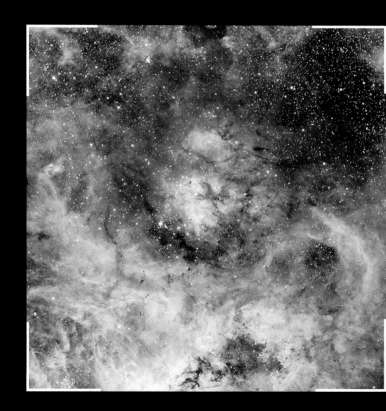

102

LMC: CLUSTER R136

At the heart of the Tarantula Nebula lies a dense star cluster called R136, packed with several dozen heavyweight blue and white stars that are just 1–2 million years old. Ultraviolet radiation from these stars powers the emission of light from across the Tarantula. The terrific rate of star formation in the Tarantula has already pushed out several older generations of stars, including a nearby cluster, Hodge 301, that is almost as impressive. In total, the Tarantula contains almost half a million solar masses of material, and some astronomers have speculated that it will ultimately coalesce to form a globular cluster (see page 180).

◼ MONSTER STAR

At the heart of R136 sits a tight knot of stars known as R136a. At one time, this object was thought to be a single gargantuan star with around 2,000 times the mass of the Sun, but it's now known that no star can grow this big – it would blow itself apart even as it attempted to form. Nevertheless, the largest single star in the cluster, R136a1, is still an awesome object– the most massive and luminous star known with at least 265 times the mass of the Sun and 10 million times its brilliance.

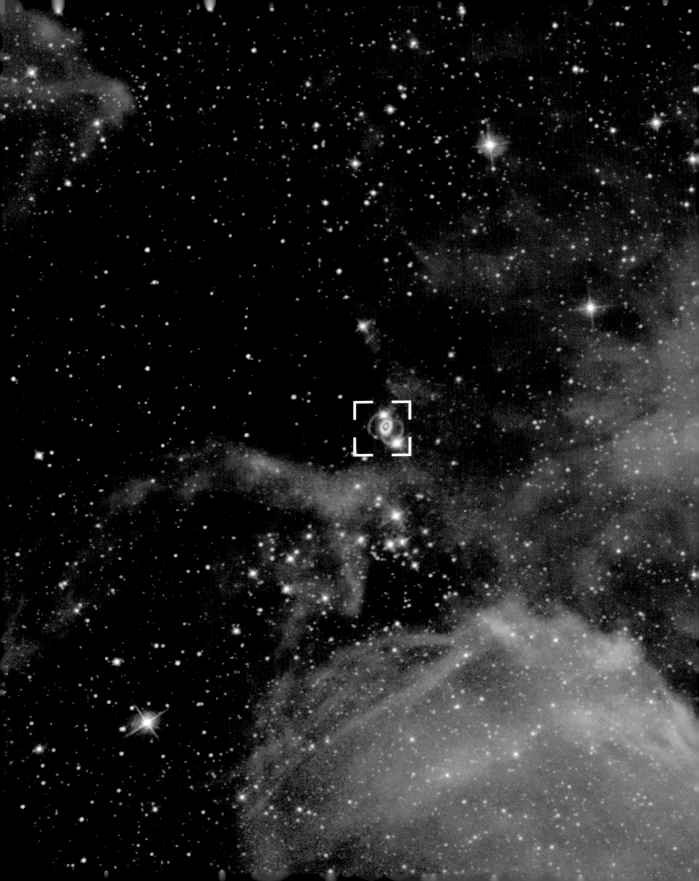

103

LMC: SUPERNOVA 1987A

On 23 February 1987, the light from an exploding star in the outskirts of the Tarantula Nebula finally reached Earth after a 168,000-year journey. As the star flared into sight, it became recognized as the closest supernova explosion to Earth since the 18th century, and was named Supernova 1987A. At its peak in May, SN1987A became a moderately bright naked-eye star, before slowly fading away over the following months. The supernova's 'progenitor' star was eventually identified as an unstable blue supergiant, which astronomers believe arose from the merger of two smaller stars some 20,000 years before.

■ EXPANDING WRECKAGE

Over the years since the initial explosion, astronomers have traced the development of SN1987A's supernova remnant, as revealed in this Hubble Space Telescope image. The expanding ring of gradually fading matter occasionally flares into brilliance again as it collides with and heats up matter in the Tarantula Nebula. Despite repeated studies, astronomers have been unable to spot the neutron star that should have been left behind by a supernova of this mass. One possibility is that the neutron star is simply hidden from view by intervening clouds of dust, but another theory is that the neutron star somehow collapsed further, creating a black hole or some even more exotic object.

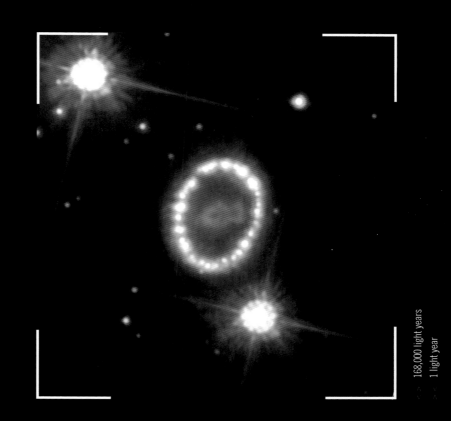

↗ 168,000 light years
↗ 1 light year

LMC: NGC 2074

The Tarantula Nebula is not the only centre of star formation within the Large Magellanic Cloud – in fact, the entire galaxy is alive with the fires of starbirth – as revealed by the Spitzer Space Telescope infrared image opposite. At the centre of another bright star-forming region lies the nebula and star cluster NGC 2074. While gravitational forces may help trigger the LMC's widespread 'starbursts', they are also self-perpetuating – as the most massive stars rapidly age and detonate into supernovae, their shockwaves initiate further waves of star formation.

▧ STELLAR NURSERY

A Hubble Space Telescope image of NGC 2074's central region reveals star-forming processes still in action today. Dark tendrils of dust-rich gas silhouetted against the glowing background gases are 'pillars of creation' similar to those seen in our own galaxy's Eagle Nebula (see pages 160–165), within which stars are forming as slowly condensing knots of matter. The red colours in this image indicate the presence of sulphur, while green originates in hydrogen atoms and blue in oxygen.

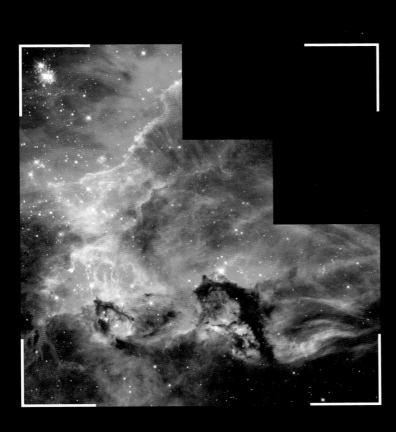

◇ 170,000 light years
⋈ 120 light years

SMALL MAGELLANIC CLOUD

The Small Magellanic Cloud (SMC) lies close to the LMC in Earth's southern skies, but is smaller, fainter and slightly more distant. Although its structure is looser than the LMC, this smaller galaxy still contains a fairly well-defined central bar, and it too may be a small disrupted spiral. The SMC played a key role in our growing understanding of cosmic distances – by assuming that all its stars were at around the same distance, astronomers were able to identify a group of pulsating stars called Cepheids, whose pulsation period is linked to their average brightness and which can therefore be used to work out the distance to other galaxies where they appear.

■ NGC 265

The SMC is studded with open clusters of stars, of which NGC 265 is one of the most prominent. Because the stars within such clusters are all at the same distance from Earth, they show the direct relationships between luminosity and colour revealed by the Hertzsprung–Russell diagram (see page 143). Most of the clusters brightest stars are hot blue stellar heavyweights, with slightly cooler white-hot stars generally appearing fainter. Less massive, cooler yellow and red stars are too faint to see in this Hubble image – those brilliant red stars that are visible are among the most massive of all, having raced through their life cycle and already developed into red giants.

→ 200,000 light years

↗ 65 light years

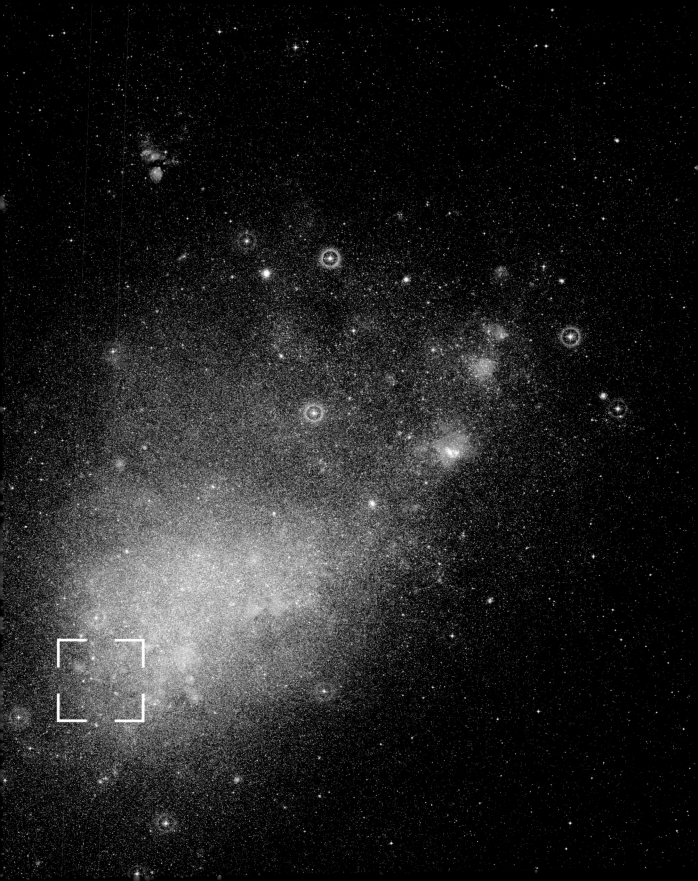

Resembling a mysterious cavern sparkling with precious stones, the beautiful star cluster NGC 602 and its surrounding nebula N90 form a spectacular treasure of the SMC. With a diameter of around 180 light years, this large and complex region highlights the processes involved in star formation. The interior of the 'cavern' has been hollowed out from within by radiation from the hot blue stars of the central cluster. This area of the sky also beautifully illustrates the three-dimensional nature of our Universe – at lower left of the image opposite, a beautiful and far more distant spiral galaxy is visible through wisps of outlying gas.

HOLLOWING OUT

Like stalactites hanging from the roof of this celestial cavern, the inner edge of N90's expanding bubble is covered by a landscape of star-forming pillars and trunks. Each of these marks a region of denser gas and dust that is better able to withstand pressure from the torrent of radiation emerging from the NGC 602 cluster. Infrared studies by the Spitzer Space Telescope have identified the traces of newborn stars warming these trunks from within.

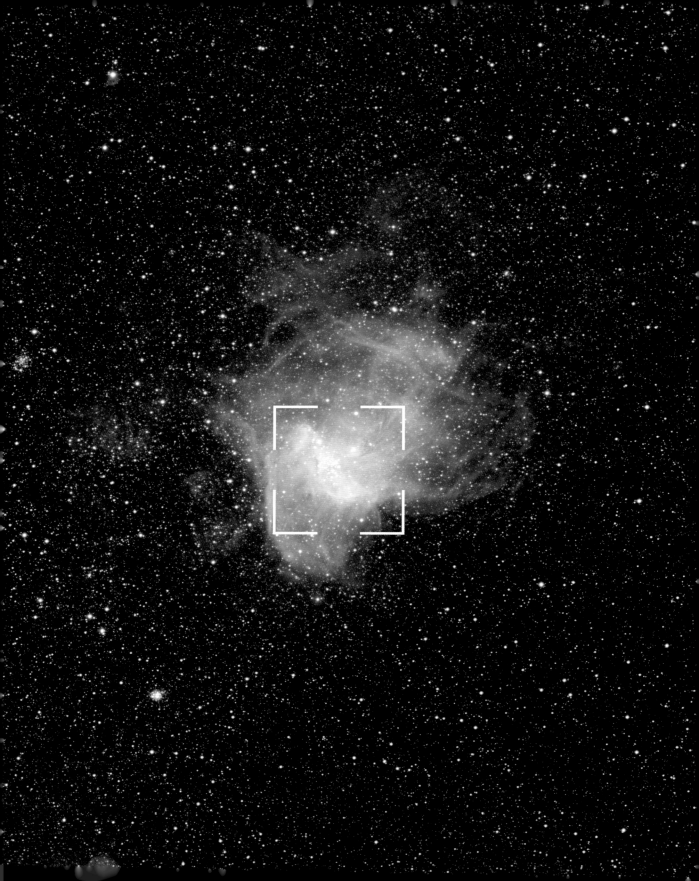

SMC: N66 NEBULA

The largest star-forming region in the SMC is N66, a cobweb-like nebula roughly 500 light years across, captured in the image opposite, from the European Southern Observatory's 2.2-m (7-ft) telescope at La Silla, Chile. Star formation has been going on in this part of the SMC for longer than in NGC 602, and a rich open cluster called NGC 346 is scattered across much of the nebula. Radiation from these stars has blown much of the nebula's gas out across surrounding space, but star formation continues near the centre.

■ WIND-SCULPTED BEAUTY
A Hubble Space Telescope close-up of the nebula's central regions reveals intricate details sculpted by the stellar winds, as well as a new generation of faint, still-forming stars along the central dust lane. The brightest and most massive stars in the cluster were born around 5 million years ago, and the same events that created them also began the slower formation of these less massive stars. However, because less massive stars actually take longer to coalesce and ignite (due to their weaker gravity), the fainter stars are only now breaking free of their dusty cocoons and becoming visible.

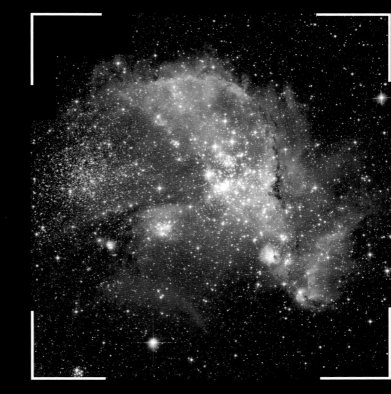

ANDROMEDA GALAXY M31

A fuzzy, star-like blob of medium brightness in the constellation of Andromeda is the most distant object visible to the naked eye – a spiral galaxy larger than our own Milky Way, and 2.5 million light years away from Earth. The Andromeda Galaxy is an enormous star system whose full extent covers three times the diameter of the Moon in Earth's skies. Although larger than the Milky Way, it is apparently somewhat less massive – together, the Milky Way and Andromeda galaxies dominate a small cluster of several dozen galaxies, known as the Local Group.

■ A DOUBLE CORE?

The combined light of billions of stars makes Andromeda's central bulge visible to the naked eye from Earth, but also renders it a featureless blaze of light in most long-exposure photographs. This Hubble Space Telescope close-up is an exception – using special image-processing techniques, it reveals details at the very centre of Andromeda – and what appears to be a 'double nucleus' with not one but two dense knots of stars, separated by about five light years. The fainter concentration marks the true centre of the galaxy, centred on a supermassive black hole of around 30 million solar masses. The brighter star cloud may simply be a temporary 'line-of-sight' effect.

[109] ANDROMEDA: DUST LANES

As seen from the Milky Way, the Andromeda Galaxy is oriented at 13 degrees from edge-on – an angle that happens to display its spiral arms beautifully, as seen in the image opposite, from the National Optical Astronomy Observatory's Kitt Peak telescope. Knots of brilliant stars in open clusters, often associated with the pinkish hydrogen glow of star-forming nebulae, trace out regions where interstellar clouds of matter are being compressed by the galaxy's rotation. Elsewhere, similar clouds appear as dark dust lanes, silhouetted against the background glow from billions of stars.

■ SECRETS REVEALED

An infrared view from the Spitzer Space Telescope reveals the startling nature of Andromeda's 'skeleton'. In this false-colour image, red light traces the path of interstellar dust clouds that help define the star-forming regions and spiral arms. Intriguingly, there are hints that Andromeda may be developing from a spiral into a more ring-like structure. 'Holes' within the dust rings may have been caused by a collision with a smaller galaxy in Andromeda's relatively recent past.

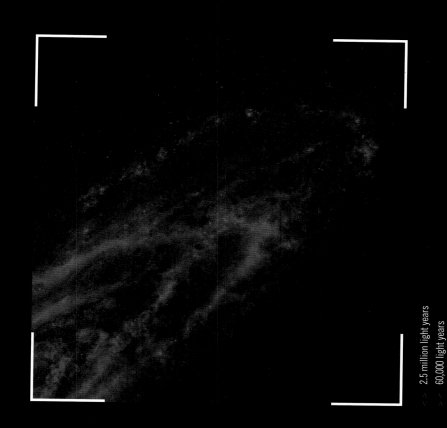

<< 2.5 million light years
>> 60,000 light years

110

TRIANGULUM GALAXY M33

Lying close to Andromeda in Earth's skies, the Triangulum Galaxy is the third major galaxy in the Local Group. However, with only half the diameter of the Milky Way it is considerably smaller and less massive than either of its spiral neighbours. Triangulum displays much more of its spiral 'face' to the Milky Way than Andromeda does, as shown in images such as the one opposite from the National Optical Astronomy Observatory at Kitt Peak, Arizona. Radio and infrared images of M33 reveal a disc of gas and dust extending well beyond the galaxy's visible star-forming regions.

NGC 604

Triangulum is sometimes described as a 'flocculent' spiral, with a relatively 'fluffy', loosely defined spiral shape (in contrast to, for example, the Pinwheel Galaxy overleaf). Nevertheless, it is home to some large star-forming regions such as NGC 604, shown here in a Hubble Space Telescope image. This tangled region, some 1,500 light years across, contains at least 200 heavyweight blue stars in a cluster that is just 3 million years old. Astronomers believe that star formation in flocculent spirals is spread by local effects – most notably the compression caused by shockwaves as previous generations of stars turn supernova.

3 million light years
1,500 light years

PINWHEEL GALAXY M101

The spectacular Pinwheel Galaxy lies well beyond our Local Group, some 25 million light years from Earth in the constellation of Ursa Major (the Great Bear). Turned face on to Earth, it is almost twice the diameter of the Milky Way, and displays an impressive and well-defined spiral structure, studded with brilliant open clusters and some 3,000 separate star-forming nebulae, and criss-crossed with dark dust lanes. Despite its size and sharp definition, the Pinwheel is far from perfect – in fact, it has been noticeably distorted by a close encounter with another galaxy at some point in its relatively recent past.

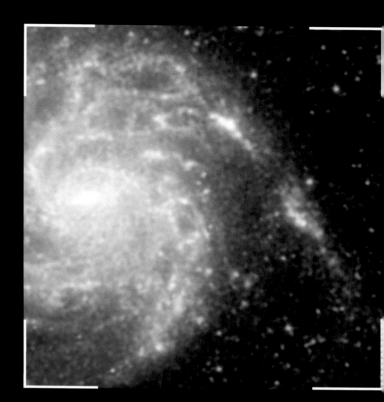

■ DUSTY SPIRAL

An infrared view of the Pinwheel from the Spitzer Space Telescope reveals different temperatures in its gas clouds associated with regions of star formation. Yellow and green colours indicate cooler regions, while pink indicates the hotter areas, which correspond to the brightest star-forming regions in visible light. Astronomers believe that the Pinwheel's current blaze of star formation has been triggered by its recent close encounter – gravitational tides raised by interaction with a smaller neighbouring galaxy have helped to amplify the spiral wave of compression that sweeps around the disc, intensifying the visible spiral pattern as new star clusters ignite.

MESSIER 74

This beautiful face-on galaxy in the constellation of Pisces (the Fish), is a magnificent example of a 'grand design' spiral galaxy, and a colourful image from the Spitzer Space Telescope (opposite) reveals its infrared structure in stunning detail. Cool gas and dust appear in red, warmer regions in yellow and green, while the hottest gas and stars appear blue. The blaze of light emerging from the core of the galaxy coincides with an X-ray source detected by other astronomy satellites, suggesting that, like our own galaxy, M74 has a supermassive black hole at its centre.

32 million light years

30,000 light years

◾ GRAND DESIGN

M74's elegant, symmetrical spiral arms are sharply defined by bright star clusters and regions of new starbirth. Slightly smaller than our Milky Way, and containing around 100 billion stars, this galaxy is thought to

encounters with one or more of its smaller neighbours (as with the Pinwheel – see previous page). However, in this case the encounters have not distorted the shape of the galaxy, merely intensified its

[113] MESSIER 66

The constellation of Leo (the Lion), is one of the most instantly recognizable of all the star patterns in Earth's skies, and the constellation is also home to a small but interesting cluster of galaxies. Known as the 'Leo Triplet', they lie beneath the lion's haunch, some 35 million light years away. All three galaxies are spirals of intermediate size, but they are all tilted at different angles to Earth, with one almost edge-on and crossed by a narrow dust lane.

■ LEO HEAVYWEIGHT

The largest of the Leo Triplet, Messier 66, is an intriguing galaxy revealed in detail by this stunning Hubble close-up. Its spiral arms are noticeably asymmetrical, and the galaxy's bright core is displaced from its centre, most likely by the gravitational pull of the other nearby galaxies. Interaction with its neighbours also seems to have intensified the rate of star formation in M66, creating brilliant open clusters and sharply defined arms.

35 million light years
65,000 light years

The spiral galaxy Messier 81 lies just beyond the limits of our Local Group at a distance of around 12 million light years in the constellation of Ursa Major. The beautiful composite image opposite combines ultraviolet data from the GALEX satellite (blue), with a Hubble Space Telescope image (white) and an infrared view from the Spitzer Space Telescope (red). It clearly shows how hot newborn stars (the major sources of ultraviolet radiation) concentrate in the spiral arms alongside the infrared glow of warm gas, while the galaxy's core is dominated by lower-mass, longer-lived and more sedate stars that shine in visible light and infrared, but not in the ultraviolet.

■ M81 CLOSE-UP

Messier 81 is one of the easiest spiral galaxies to observe on account of its compact nature and large, bright central hub, which is thought to harbour a black hole with the mass of 70 million Suns. Named Bode's Galaxy after the German astronomer Johann Elert Bode, who discovered in 1774, it lies at the centre of its own galaxy group, with at least 34 members at the latest count.

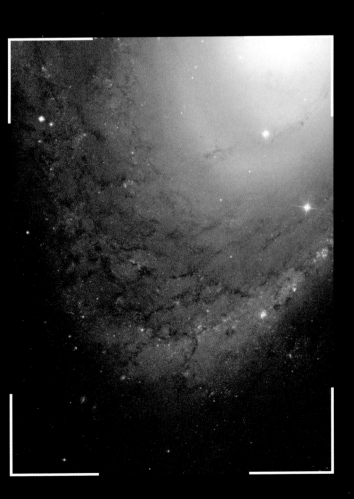

∧ 12 million light years

∨
∧ 20,000 light years
∨

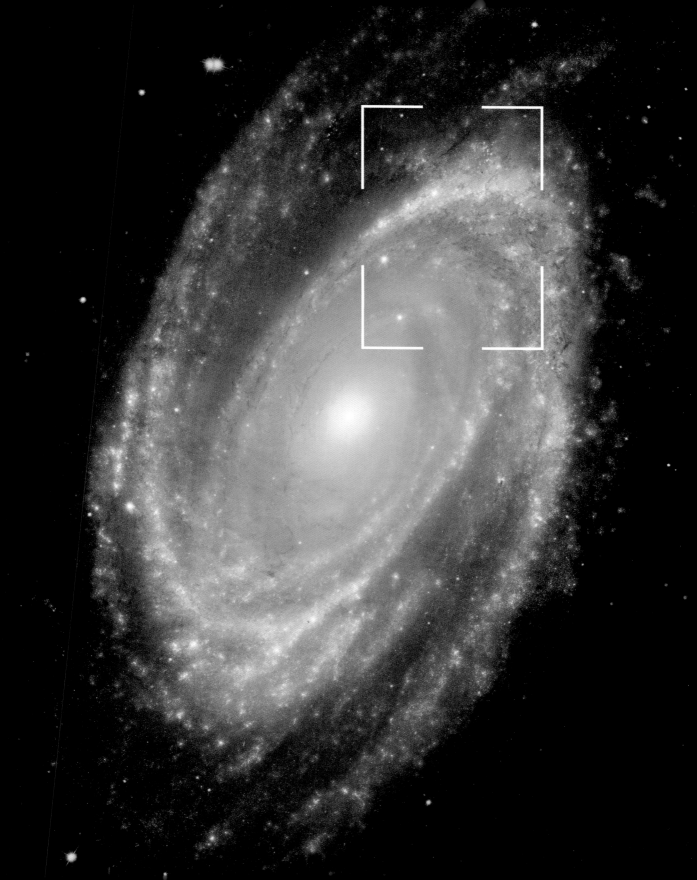

[115] CIGAR GALAXY M82

Lying close to Bode's Galaxy in the constellation Ursa Major (see previous page) as viewed from Earth, the so-called Cigar Galaxy Messier 82 is only separated from it in space by 300,000 light years. This remarkable irregular galaxy, seen edge-on and with a dark dust lane running along its centre, is alive with the fires of starbirth and surrounded by flows of gas seemingly rushing to escape it – little wonder that astronomers once described it as an 'exploding galaxy'.

BURST OF STARBIRTH

In reality, M82 is the closest and brightest example of a 'starburst galaxy' – a gas-rich irregular galaxy fired into a huge wave of starbirth by a close encounter with a larger neighbour. In many ways, M82 is suffering a much larger-scale version of the turmoil seen in our own satellite galaxy, the Large Magellanic Cloud (see page 231). The 'explosive' appearance is caused by supernova shockwaves heating gas in the nebulae around them, and accelerating it to such speeds that it escapes the galaxy's relatively weak gravity.

11.5 million light years

4,000 light years

SOUTHERN PINWHEEL M83

Magnificent Messier 83, just 15 million light years away in the constellation of Hydra (the Water Snake), is one of the finest examples of a distinct class of spiral galaxies known as 'barred spirals', in which, rather than connecting directly to the hub, the spiral arms emerge from either end of a long bar that crosses the centre of the galaxy. Barred spirals account for around two-thirds of all spiral galaxies, including our own Milky Way and probably its neighbour Andromeda. Astronomers believe that bars are temporary phenomenon that come and go throughout the life of a typical spiral.

ARC OF FIRE

This Hubble Space Telescope close-up of star formation in M83 covers a range of wavelengths from the near-infrared, through visible light to the near-ultraviolet. Along the left side, a spiral arm is studded with star-forming nebulae, open and globular clusters, expanding bubbles of supernova remnants and some particularly brilliant individual stars. Behind everything else, countless fainter stars merge together to create a billowing background glow.

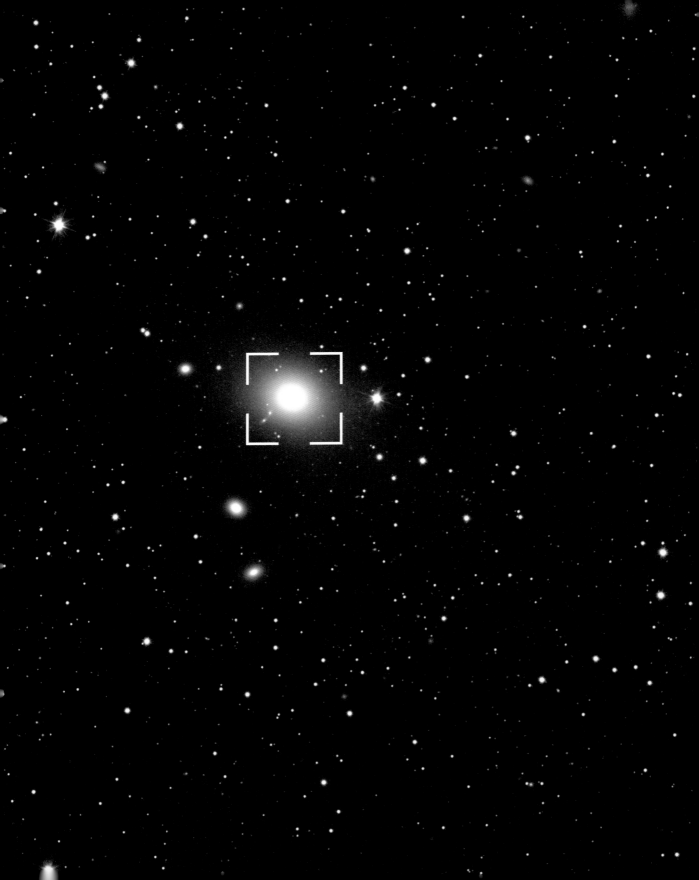

The most prominent elliptical galaxies in the neighbourhood of the Milky Way are M32 and M110, two small satellites of the Andromeda Galaxy (see page 244). Further afield, though, there are larger elliptical balls of stars, some of which can match the scale of our own spiral galaxy, and some of which dwarf it. Giant ellipticals are enormous spherical star clouds with up to 200 times the mass of the Milky Way – the largest galaxies known. The nearest such galaxy to Earth is Messier 87, around 55 million light years away at the heart of the Virgo galaxy cluster (see page 280).

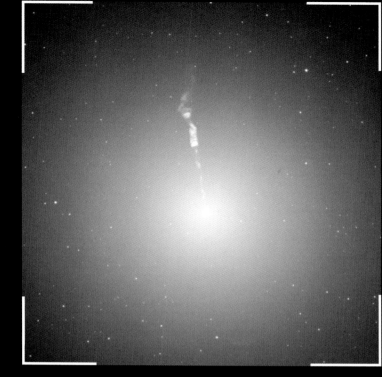

POWERFUL JET

Using image-processing techniques, the Hubble Space Telescope peers through the outer stars of M87 to focus on the bright jet that shoots out from its centre. Particles in the jet travel at close to the speed of light, and are thought to originate in the region around M87's central supermassive black hole – as matter spirals down onto the hole, some of it gets caught up in its tangled magnetic field and is ultimately fired out from the poles at incredibly high speeds. This is just one, relatively mild, example of a phenomenon known as an 'Active Galactic Nucleus'.

The strange galaxy NGC 5128, some 11.5 million light years away, is the brightest and closest major active galaxy to Earth, and shines at just below naked-eye visibility. Better known as 'Centaurus A', from the designation of the powerful radio source at its heart, this 'lenticular' ball of stars has a dense lane of dust running across its centre. Multi-wavelength composites such as the image opposite (which shows radio waves in orange and X-ray emissions in blue) reveal two huge jets bursting from the centre and billowing out into clouds that flank the visible galaxy.

DUSTY HEART

Centaurus A is the most prominent active galaxy in Earth's skies, and appears to be the result of a recent collision and merger between a large elliptical galaxy and a smaller spiral. The 'ghost' of the cannibalized spiral is still visible in the form of the dark dust lane, which can be traced deep within the galaxy. Near the centre, this Hubble infrared image reveals the glow from a disc of hot gas spiralling in towards Centaurus A's central supermassive black hole. Astronomers believe such galactic collisions play a crucial part in triggering bursts of activity.

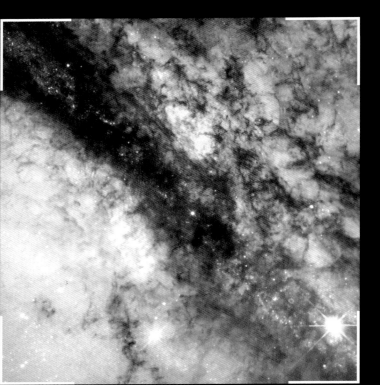

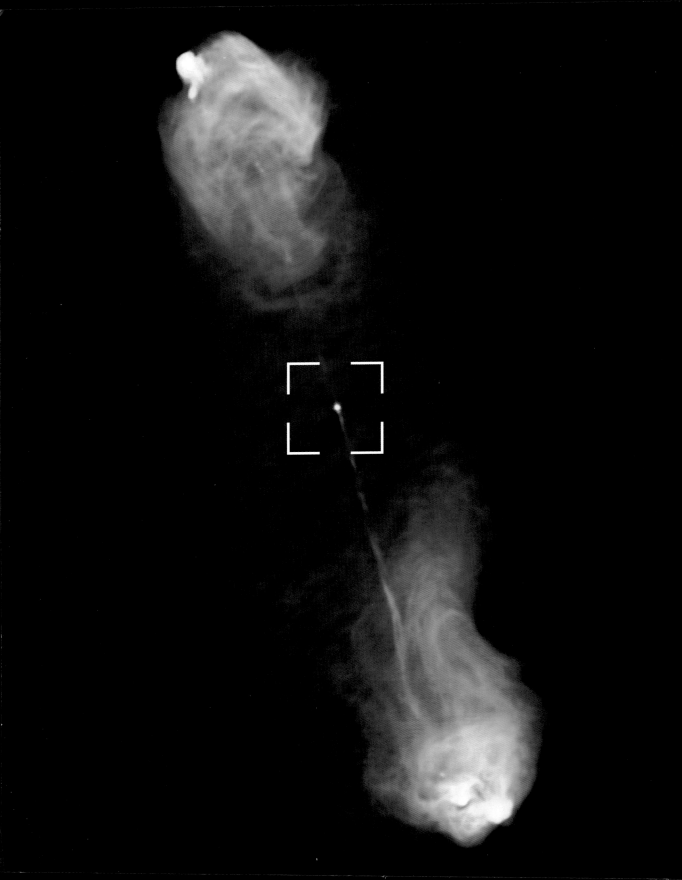

When astronomers developed the first radio telescopes in the 1930s, they discovered a number of bright radio sources in the sky, including a strong one in the constellation of Cygnus (the Swan). Modern radio images show that Cygnus A is actually a pair of glowing 'lobes' with a fairly unremarkable-looking galaxy between them. Cygnus A is now known to be a classic example of a 'radio galaxy' – an active galaxy in which the nucleus region lies edge-on to Earth, and only the radio glow created by its escaping jets reveals anything unusual. It is one of the largest objects in the sky, with an overall diameter of half a million light years.

■ CENTRAL GALAXY

This processed image from the Keck II Telescope on Mauna Kea, Hawaii, reveals the galaxy that lies between the twin lobes of Cygnus A as a misshapen blaze of light. Colours from dark blue through to red and yellow indicate increasing brightness levels, with the brightest spots of all 'burnt-out' in black. The structure of the galaxy is hard to interpret, but the peculiar shape and thick dust lanes running across the centre suggest that it is in fact a galaxy merger that has sparked one of the central black holes into life.

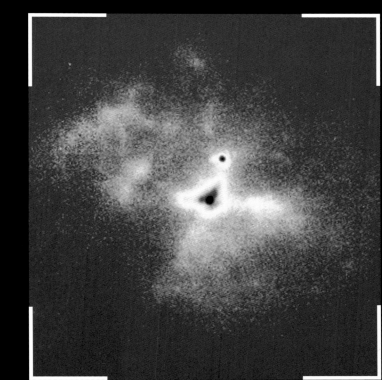

120

QUASAR 3C 321

Quasars are the most violent form of active galaxy – so bright but distant that, when they were first discovered in the 1950s, astronomers actually thought they were a strange new class of star. It was only when careful observations revealed the 'host galaxies' surrounding their bright nuclei that it became clear they were stranger than anyone had imagined. These remarkable images highlight an unusual and relatively nearby quasar pair. Opposite, X-ray and radio data (shown in purple and blue) reveal a stream of particles some 850,000 light years long, blasting out of one galaxy and straight through its neighbour.

DEATH STAR QUASAR

A close-up, visible-light image from the Hubble Space Telescope reveals the brighter of the two galaxies in the 3C 321 pairing, roughly 1.4 billion light years from Earth. Quasars are a far more powerful version of a Seyfert galaxy (see page 271), in which the black hole at the centre of the galactic nucleus is feeding on its surroundings at such a rate that the hot 'accretion disc' around it becomes large and bright enough to outshine the rest of the galaxy completely.

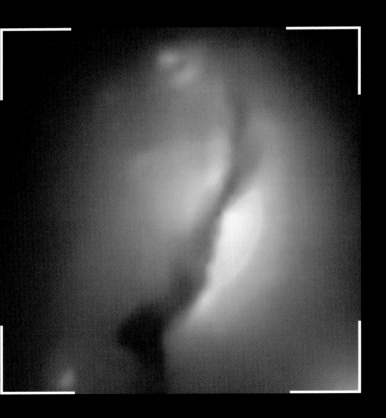

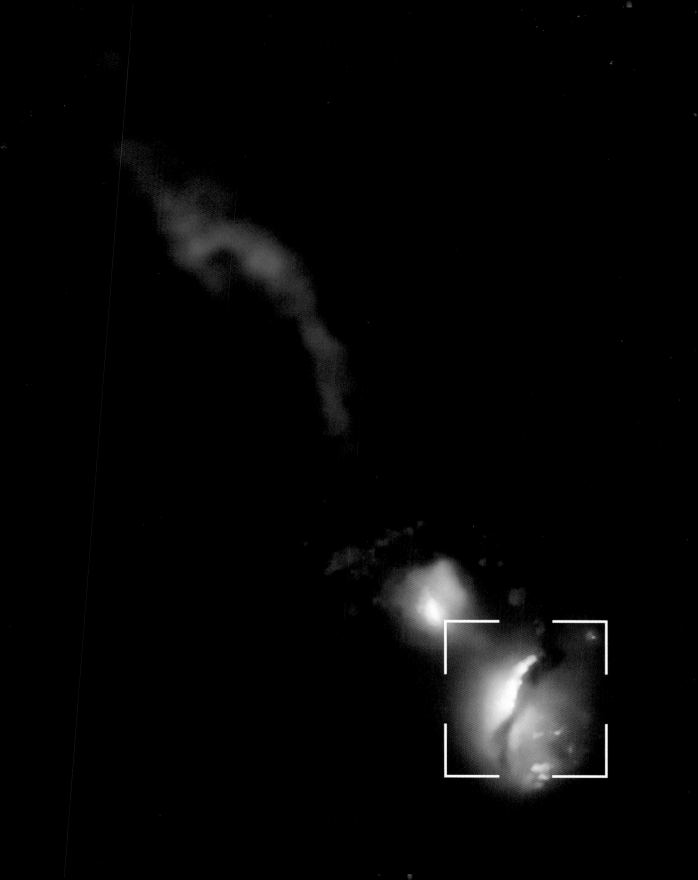

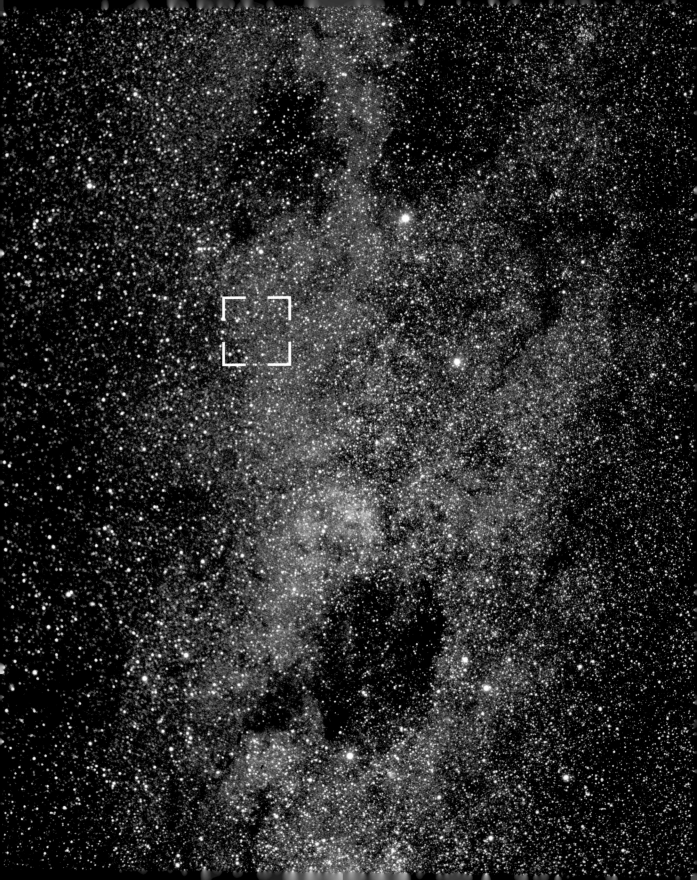

121

ESO 97 G-13

Despite being just 13 million light years from Earth, one of the closest active galaxies was not discovered until the 1970s. Known as the Circinus Galaxy, this small spiral stayed hidden for so long because it happens to lie behind a dense patch of the Milky Way. It is an example of a 'Seyfert' galaxy – a spiral with a moderately active nucleus, whose core appears far brighter than would normally be expected – and is prone to changes in brightness on a timescale of hours to days.

COLOURFUL CORE
This Hubble image of the Circinus Galaxy's central region shows glowing hot gas (pink) ejected from the nucleus, as well as two broad doughnut-shaped rings of dust surrounding it. The bright spot at the centre is the active galactic nucleus itself – a blazing disc of gas and dust that is heated to incredible temperatures and emits a wide range of wavelengths as it is pulled to its doom in the central black hole.

13 million light years
26,000 light years

COLOUR OF COLLISION

Although collisions between individual stars are rare during galactic mergers, the sparse clouds of gas and dust that form a galaxy's skeleton run straight into one another, creating enormous shockwaves that trigger huge bursts of star formation. In this colourful composite image, infrared (red) and X-ray (blue) images are overlaid onto a visible-light Hubble image, revealing the infrared glow of star-forming regions and a surrounding haze of superhot X-ray gas that has been driven away from the galaxies by their collision.

45 million light years
20,000 light years

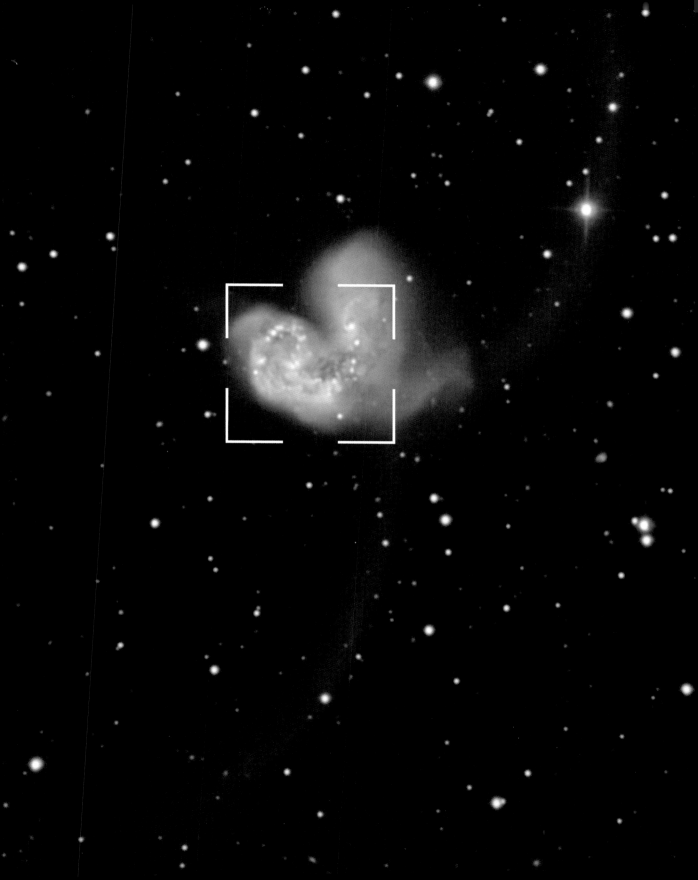

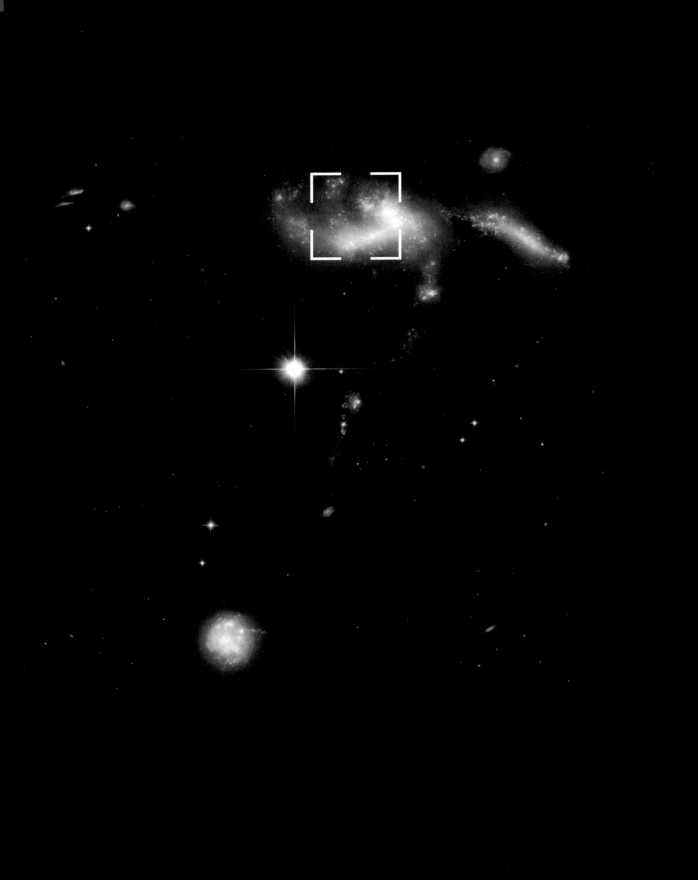

HICKSON COMPACT GROUP 31

Not all galactic mergers are as spectacular as the collisions of the Antennae or the Mice (see pages 272 and 276). The four dwarf irregular galaxies of Hickson Compact Group 31 (HCG 31) are engaged in their own slow gravitational waltz with less fire and drama, but their story is just as intriguing. Astronomers believe that dwarf galaxies such as these are the building blocks of larger galaxies – as these gas-rich star clouds come together they eventually grow large enough to develop spiral structures.

FOSSIL GALAXIES

One of the strangest things about HCG 31 is the group's relative proximity to Earth. At a distance of 'just' 166 million light years in the constellation of Eridanus (the River), these galaxies are on our doorstep in cosmological terms. In stark contrast, most other merging dwarf galaxies are many billions of light years away (see page 303). Based on this finding, astronomers had believed that all galaxies of this type had merged into larger systems long ago – finding HCG 31 so close to Earth is rather like finding a 'living fossil'.

166 million light years
30,000 light years

Another well-known pair of interacting galaxies share a catalogue number of NGC 4676. The 'Mice', situated around 290 million light years away in the constellation of Coma Berenices, are named on account of the long 'tails' of stars that trail away from them in opposite directions. These enormous streams of stars formed as the galaxies approached each other prior to a close encounter 160 million years ago. Tidal forces caused by each galaxy's gravity pulling on the near and far sides of its neighbour with different strengths, ultimately caused the spiral arms to unwind across space.

DUSTY CORES

A Hubble Space Telescope close-up of the twin galaxies reveals stunning details, including broad lanes of disrupted dust thrown across the space between the two galactic hubs. Although currently moving away from each other, the two galaxies are now gravitationally bound and their fate is sealed – in another few hundred million years, they will have merged completely. Five billion years from now, a similar fate awaits the Milky Way and Andromeda galaxies of our own Local Group.

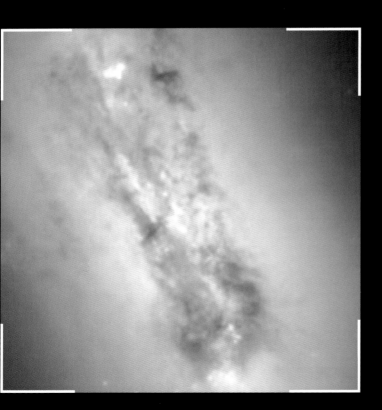

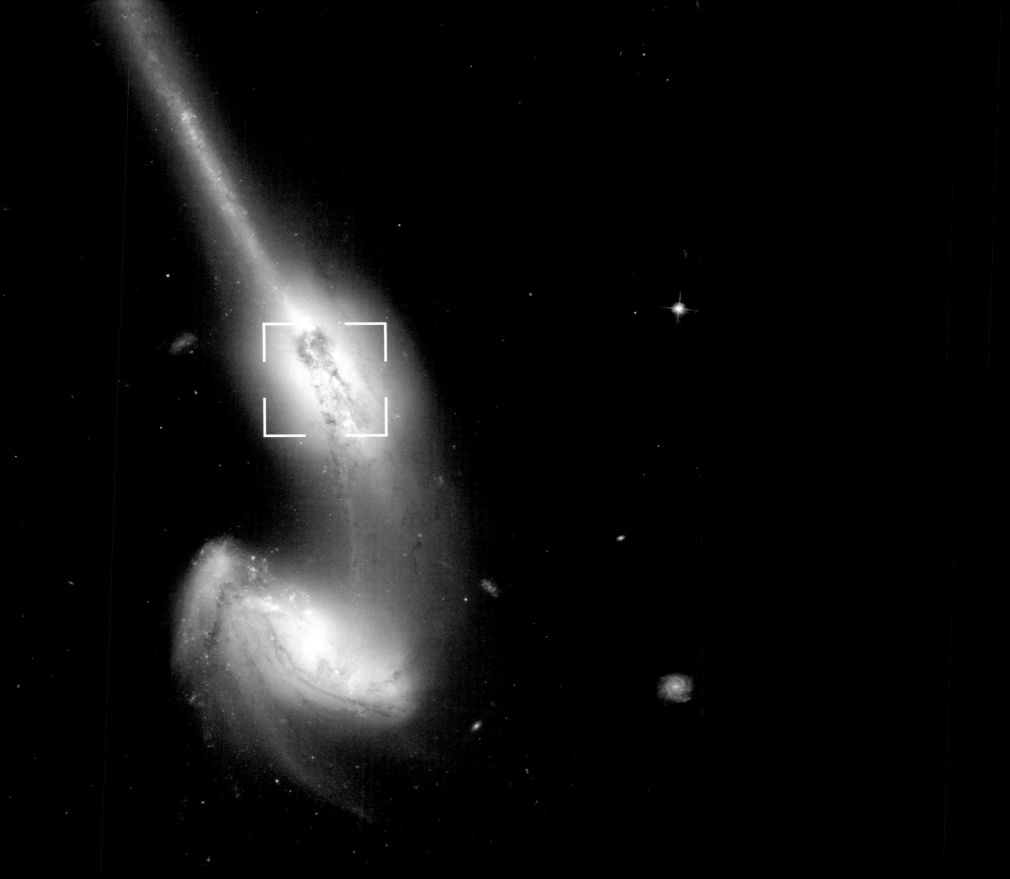

STEPHAN'S QUINTET

In 1877, French astronomer Edouard Stephan discovered a tight cluster of galaxies in the constellation of Pegasus – the first 'compact galaxy group' to be discovered. Today we know that 'Stephan's Quintet' is actually a quartet with a fifth galaxy (the bluish NGC 7320) intruding in the foreground. The four background galaxies include a barred spiral, an elliptical and an interacting pair that were also once spirals. The visible-light and near-infrared image opposite reveals a host of faint stars in all the galaxies that are hidden from normal observations.

■ INFRARED SHOCK

A Spitzer Space Telescope image peers far deeper into the infrared than Hubble, uncovering hitherto invisible features including a spectacular shockwave (the central green arc) larger than the Milky Way. Cooler regions in this image are shown in red and warmer ones in blue, while the green indicates light emitted by hydrogen gas as it is energized by the collision between NGC 7318b, one of the interacting galaxy pair, and an unseen cloud of superhot intergalactic matter.

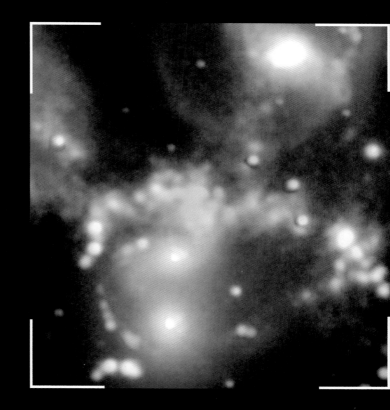

The powerful gravity of large galaxies gives them a natural tendency to form into clusters – our Local Group is a modest example, but galaxy groups can grow much larger and more densely packed. The nearest major galaxy cluster is around 59 million light years away in the constellation of Virgo (the Virgin). Although it covers a volume of space roughly equivalent to the Local Group, it contains more than 1,300 galaxies, some of which are shown in the image opposite. The Virgo Cluster forms the centre of a 'Local Supercluster' that incorporates several other clusters including our own.

■ ELLIPTICAL ANCHORS

At the heart of the Virgo Cluster lie several huge 'giant elliptical' balls of stars, including M84, M86 (shown in detail here) and M87, the largest of all (see page 263). Such galaxies are only found near the centre of large clusters, and form anchors that draw in the smaller galaxies around them, leading to spectacular collisions and mergers. This stunning image from the National Optical Astronomy Observatory reveals filaments of hydrogen trailing from the disturbed spiral galaxy NGC 4438 after a recent close encounter with M86.

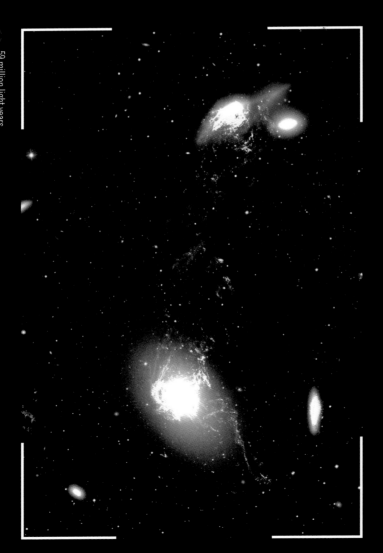

> < 59 million light years
> < 400,000 light years

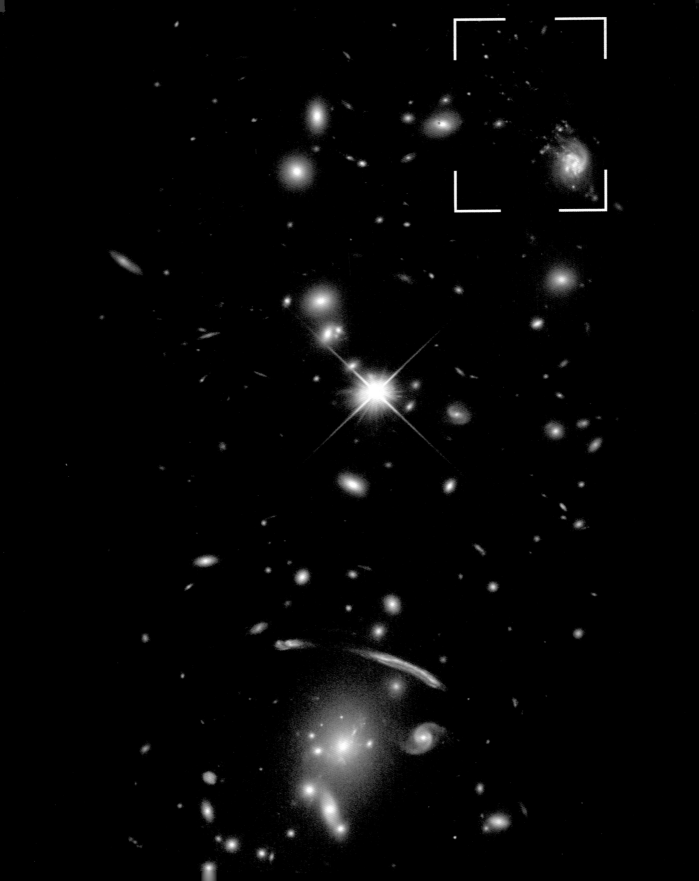

127

ABELL 2667

Cluster Abell 2667 is a huge cloud of galaxies, some 3.2 billion light years from Earth in the constellation of Sculptor. It is rich in elliptical galaxies floating in a 'sea' of superhot multi-million-degree gas stripped away as they formed. The cluster's huge mass, equivalent to at least 350 Milky Ways, allows it to exert a powerful influence on its surroundings – it has disrupted the path of a passing spiral galaxy, and elsewhere produced the telltale streaks of gravitational lensing (see page 291).

■ CANNIBAL CLUSTER

Nicknamed the 'Comet Galaxy' by its discoverers, this spiral galaxy is suffering the consequences of a close encounter with a major cluster. Accelerated to a speed of at least 3.5 million km/h (2.2 million miles per hour), it is falling towards the cluster's centre of gravity and slamming directly into the hot gas that lies in this region. Pressure from the gas is stripping away the galaxy's reserves of gas, leaving them trailing behind on its path through space. Meanwhile, tidal forces are also sparking a wave of star formation within the doomed galaxy.

DEPTHS OF THE UNIVERSE

Cosmology is the field of astronomy that looks at the large-scale properties of the Universe and attempts to answer crucial questions about its size and shape, origins and fate. Nevertheless, even this ambitious science is still underpinned by telescopic observations, and images of the deep cosmos can reveal some of the phenomena that cosmologists study.

When astronomers look out across the empty gulf of space towards the most distant parts of the Universe, they face several challenges, but one of the most fundamental is the limited speed of light. In Earth's near neighbourhood, light's limited speed (despite being a staggering 300,000 km/s or 186,000 miles per second) causes practical issues when communicating with and controlling space probes. However, when looking over longer distances, we must sometimes bear in mind that the light now reaching Earth started its journey a long time ago.

In the local Universe, this so-called 'lookback time' effect is intriguing, but makes little difference to the value of observations themselves – galaxies remain essentially the same over tens of millions of years. However, when we look at galaxies that are billions of light years from Earth, we are seeing objects at a significantly earlier stage in their evolution – the most distant galaxies of all are also

the youngest, and it is little wonder they are dominated by bright but shapeless irregulars – the building blocks of today's more orderly galaxies.

But how do astronomers measure the distance to galaxies in the first place? The most important methods use 'standard candles' – stars whose true luminosity can be calculated independently, and whose apparent brightness can therefore reveal their true distance from Earth. In the nearby Universe, the best standard candles are Cepheid variable stars – pulsating yellow supergiants that can be spotted in galaxies up to around 100 million light years away.

Fortunately, the first measurements of galactic distances carried out in this way revealed another very useful pattern. When astronomers split the light from these galaxies into a spectrum to study their chemical 'fingerprints', they at first found it impossible to relate the dark absorption lines they saw to familiar chemicals. Then they realized that

BIG BANG

This schematic diagram summarizes the early history of the Universe. The Big Bang was followed almost immediately by a period of sudden expansion, known as Inflation, and the gradual formation of subatomic particles that coalesced into atoms as the Universe cooled. Initially, the Universe was 'foggy', but once its density fell sufficiently, light raced away from matter and the Universe entered a dark age. This only ended as the first generation of stars began to shine, and the massive black holes they left behind acted as seeds for the formation of galaxies.

they were actually seeing familiar groups of lines, displaced some way towards the red end of the spectrum from their usual positions. Different galaxies displayed different 'red shifts' in their spectral lines, and the best explanation for this strange phenomenon was that the red shift was a 'Doppler effect' – a stretching of light waves caused by the galaxy and its Earth-based observers moving apart at high speed. (In everyday life, the Doppler effect is most familiar from the shift in a siren's pitch as an emergency vehicle rushes towards us, passes us and then moves away).

The most surprising thing about these red shifts, though, was that they were closely linked to the distance of other galaxies – the further away a galaxy was, the faster (on average) it is moving away from us. The only logical way to interpret this extraordinary pattern is that the entire Universe is expanding at a uniform rate and carrying the galaxies with it.

Red shift therefore makes a useful proxy for distance measurements when a galaxy is too far

away to identify Cepheid variables within it. As a general rule, the further away a galaxy is, the greater its red shift – and indeed astronomers often describe the distance to galaxies in terms of red shifts rather than light years.

But the expansion of the Universe also has far deeper implications for cosmologists. If space is expanding and the cosmos is getting bigger, it follows that it is expanding from a smaller state, and that at some point in its distant past, everything occupied essentially the same point in the Universe, before some violent event released enough energy to begin the process of expansion. And just like the compressed air inside a bicycle pump, the Universe must have been unimaginably hotter when it was all compressed into a tiny volume. This is the origin of

This schematic shows the possible fates of the Universe. Its current steady expansion may continue at more or less the same rate, resulting in an ever larger, colder cosmos, or it may slow to a halt but never quite collapse. Alternatively, if the Universe was dense enough, gravity might pull it back to a 'Big Crunch' (though this is now thought to be unlikely). One other possibility is that, if the strength of dark energy increases, it might eventually be enough to tear matter apart completely, in a so-called 'Big Rip'.

the Big Bang Theory – the now-standard model of cosmic origins in which the Universe and all the matter within it was born in a cataclysmic release of energy now estimated to have happened 13.7 billion years ago.

Although the Big Bang offers our best model of the way in which the Universe and the matter within it were created, it raises some intriguing problems. It predicts a far more massive Universe than the one we can detect with our telescopes, and this, along with the behaviour of galaxies and galaxy clusters, is strong evidence that 80 per cent of all material in the cosmos is so-called 'dark matter' (a misnomer, since this mysterious stuff is not just dark but also apparently transparent).

And while astronomers are still puzzling over the precise nature of dark matter, recent discoveries have presented them with an even greater challenge. By using exploding stars called Type 1a supernovae, which always release the same amount of energy, as 'standard candles' for independently measuring the distance to remote galaxies, astronomers hoped to work out the rate at which cosmic expansion was slowing down. In fact, they found evidence that the rate of expansion is increasing and space is being stretched apart by a mysterious force called 'dark energy'.

Ultimately, perhaps the most amazing thing about cosmic expansion is that it puts an absolute limit on our observations of the Universe. The further across space, and the further back in time, we look, the closer the speed of receding objects approaches the speed of light. What happens when we reach that limit? While it's perfectly possible for the physical Universe to stretch beyond that, no light will ever reach us from such remote realms – it has not yet had time to reach us in the 13.7 billion years since the Big Bang, and thanks to the expansion of space, it never will. Instead, we run into a wall – the edge of our 'observable Universe'. As our largest telescopes sweep up light from ever-fainter objects and get closer and closer to this point, we may see the very first generations of giant stars that are thought to have formed the seeds of galaxies themselves – but for the moment, our only view of this great cosmic boundary comes in the radio waves of the Cosmic Microwave Background – a softly glowing echo left behind by the incandescent fireball of the Big Bang itself

128

SDSSJ0946+1006

This bizarre-looking distant object, named SDSSJ0946+1006 and captured for the first time by the Hubble Space Telescope, is an Einstein ring —conclusive proof that Einstein's theory of general relativity is correct. The theory explains how large masses distort space and time around them – a model that can describe the effects of gravity with great precision. One significant effect, displayed in phenomena such as the Einstein ring, is called 'gravitational lensing' – it explains how large masses affect the path of light rays which (since they have no mass) should be immune to the effects of gravity.

■ EINSTEIN RING

A close-up of the Einstein ring reveals further detail including a faint secondary ring. The object at the centre of the ring is a large elliptical galaxy roughly 3 billion light years away. Directly behind it lie two more galaxies at distances of 6 billion and roughly 11 billion light years from Earth. Normally, the distant galaxies would be completely hidden from our view, but as their light rays pass by the foreground elliptical, the distortion of space itself bends the path, focusing the light back towards Earth and creating the ring-like effect. Astronomers believe the odds against such a triple alignment are 10,000 to 1.

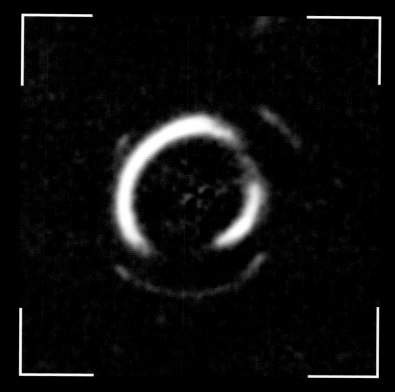

>> 3.2 billion light years
>> N/A

Although rarely as orderly as the Einstein ring on the previous page, gravitational lensing is fairly common on cosmic scales. Here it is caused by the enormous concentrations of mass in galaxy clusters, which (being far larger than individual galaxies) typically have many galaxies lying 'behind' them from our point of view. Diverging light from these remote galaxies is warped to create bright arcs around the foreground cluster. Because lenses typically have an intensifying, as well as a distorting, effect, they make it possible to see distant, faint galaxies that would normally be beyond the range of our telescopes.

■ COSMIC ZOOM

One of the most impressive gravitational lenses so far discovered is the galaxy cluster Abell 1689 in the constellation of Virgo. Lying roughly 2.2 billion light years from Earth, it contains hundreds of massive galaxies paced into a region of space about 2 million light years across, and so acts as a powerful lens, creating the bright blue-white arcs seen between the yellowish foreground galaxies. This composite image, incorporating observations from the Chandra X-ray Observatory, reveals the hot gas between the visible galaxies, which also contributes to the lensing effect.

DARK MATTER

Gravitational lensing effects also provide an important way for astronomers to probe the distribution of the mysterious 'dark matter' that outweighs normal matter in the Universe by a factor of about four to one (see page 289), and which can only be detected by its gravitational effects. By analysing the strength and pattern of lensed images created by a particular cluster, it's possible to reconstruct the 'shape' of the distorted space creating the lensing effect, and therefore the distribution of mass within the foreground cluster.

DARKNESS VISIBLE

The 'Bullet Cluster' IE 0657-556 formed from the head-on collision of two clusters 150 million years ago. While the individual galaxies have mostly passed straight by one another as revealed by the visible-light Hubble image opposite, X-ray observations from Chandra (shown at left in pink) reveal that their hot gas has collided head-on, creating a distinctive bullet-shaped shockwave. The gas in these clouds weighs far more than the visible component of the galaxies themselves, but a map of the cluster's mass derived from its lensing effects shows that despite the collision, the undetectable dark matter (blue) is still concentrated around the galaxies themselves.

3.8 billion light years

1.5 million light years

MESSIER 100

This spiral galaxy in the Virgo Cluster (see page 280) was an early target for the Hubble Space Telescope's Cepheid hunt. In total, some 20 separate Cepheid variables were identified and measured within it (one of which is shown circled), yielding a distance of 55 million light years. By comparing this and other similar measurements to the red shifts in each galaxy's light, the 'Hubble Key Project' team worked out that the Universe is expanding at a rate of 22.7 km/s (14.1 miles per second) per million light years. This value for the 'Hubble Constant' can now be used to estimate the distance of more remote galaxies based on their red shifts alone.

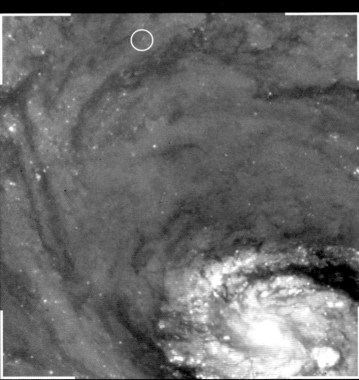

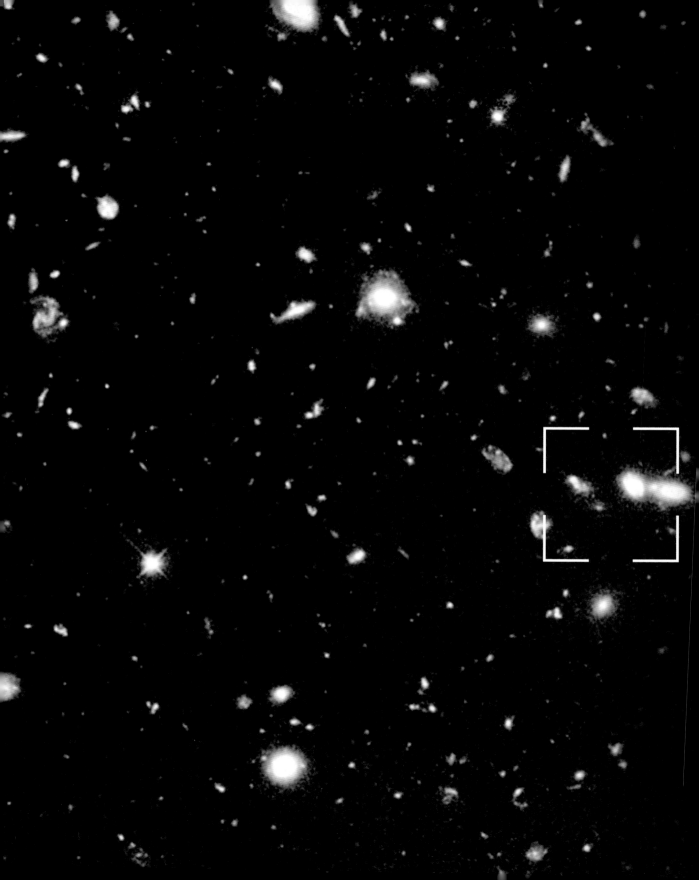

DISCOVERING DARK ENERGY

Just as the Hubble Space Telescope has helped to clarify some of the biggest questions about the size, expansion and age of the cosmos, new ones have been emerging. Observations of distant 'Type 1a' supernovae have suggested that the Universe is expanding more rapidly than can be explained through traditional models, driven by a mysterious force called 'dark energy'. According to the latest estimates, dark energy may account for 74 per cent of all the 'stuff' in our Universe, with dark matter accounting for a further 22 per cent, and visible matter just 4 per cent.

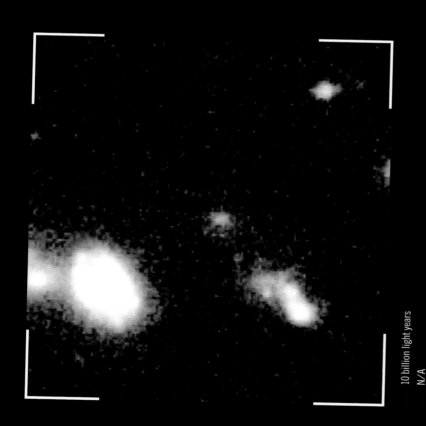

10 billion light years
N/A

SUPERNOVA 1997J

Type 1a supernovae are a special type of supernova created in rare situations where a white dwarf (see page 211) accumulates new matter onto its surface and collapses suddenly into a neutron star (see page 218). These supernovae always brighten and fade in a distinctive way, and because the collapse always releases the same amount of energy, they always peak at the same luminosity. This makes them useful 'standard candles' for working out the distance of their host galaxies – especially because they can be seen over billions of light years, as revealed in this image of a Type 1a supernova 10 billion light years from Earth.

In 1996, astronomers turned the powerful gaze of the Hubble Space Telescope onto a small patch of sky in the constellation Ursa Major. By observing the same small region for a total of ten days, they produced the deepest view of the Universe up until that time – the 'Hubble Deep Field'. The results were astounding – a sky filled with at least 1,500 galaxies stretching off to the very limits of visibility, billions of light years away in time and therefore billions of years back in history.

» 11 billion light years

» N/A

■ DEEP DETAIL

A detailed sample from the Hubble Deep Field reveals a bewildering variety of galaxies, including relatively large and well-developed spirals and ellipticals, but also an array of more distant, smaller and shapeless blobs. In general, the 'red shifting' of light from more distant, faster-receding galaxies (see page 288) should make more distant galaxies appear redder, but in fact, there is a pronounced 'blue excess' in the smaller galaxies. This is probably because these remote galaxies are still in their infancy and most of their light comes from brilliant blue and white stars that have recently formed.

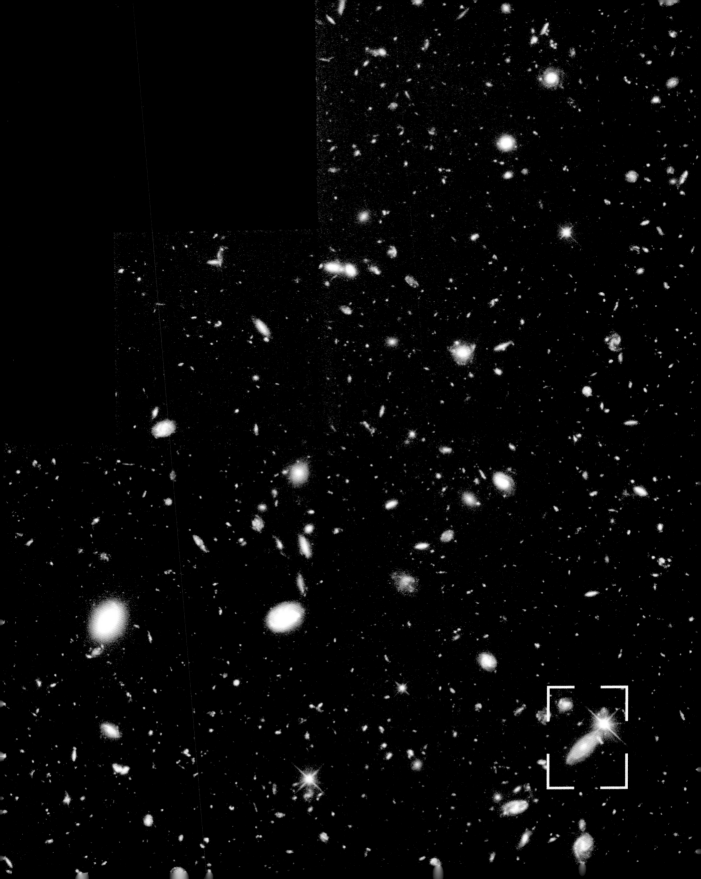

HUBBLE ULTRA-DEEP FIELD

The success of the Hubble Deep Field experiments inspired several successors, including the so-called 'Hubble Ultra-Deep Field' (HUDF), which used an upgraded Hubble camera to stare at a small area of the southern constellation of Fornax (the Furnace) for almost a million seconds in late 2003. The HUDF captured images of 10,000 galaxies up to 13 billion light years away – so distant that their light must have left on its journey just 700 million years after the Big Bang itself.

■ GALAXY ZOO

Light from some of the most distant galaxies is red shifted to such extremes that it arrives at Earth as infrared waves, and so the HUDF was complemented by a similar survey from Hubble's infrared camera. This revealed the most distant galaxies of all, still in the early stages of formation but alive with the fires of rapid starbirth. It confirmed that, 1 billion years after the Big Bang, there were no large spiral or elliptical galaxies, but a large population of small, gas-rich dwarf galaxies that later have merged to form today's galaxies (though see page 275 for some rare survivors).

13 billion light years
N/A

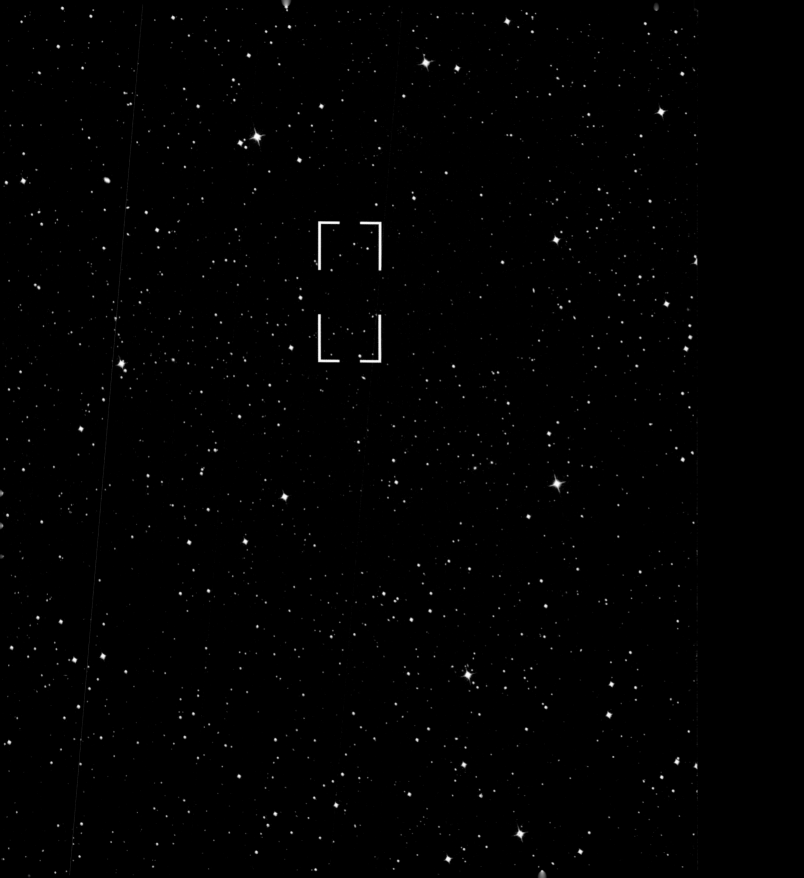

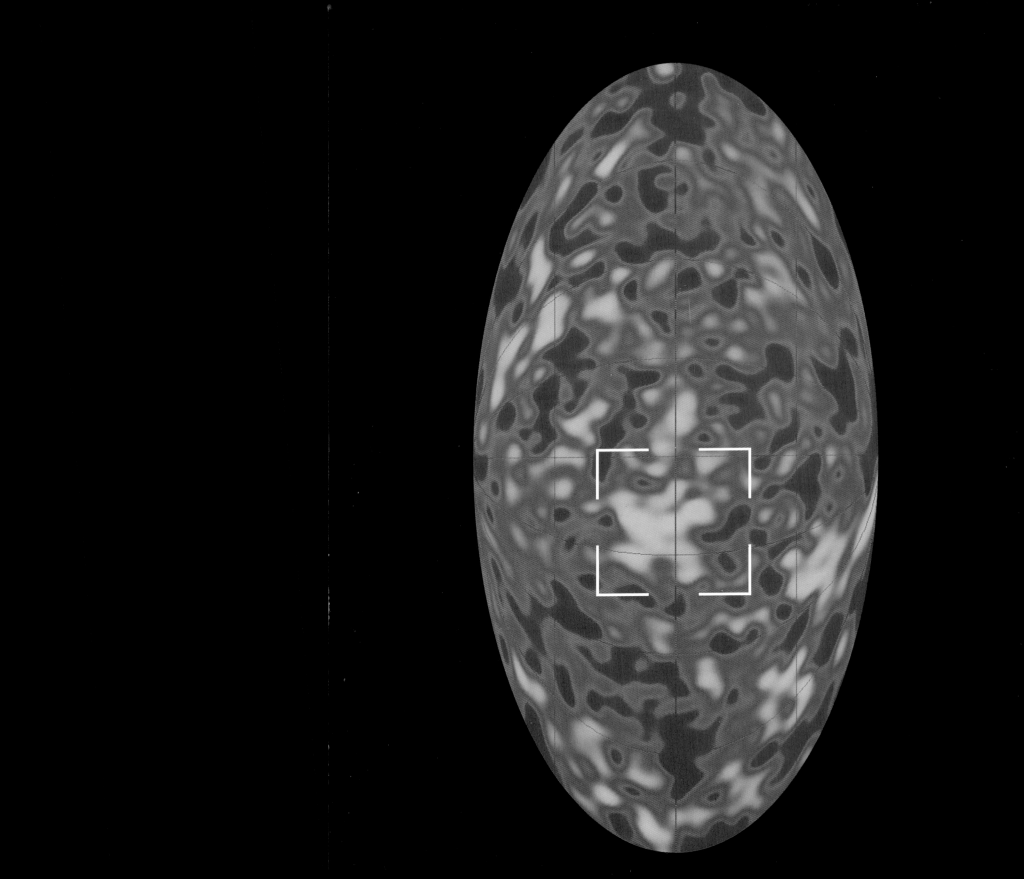

COSMIC MICROWAVE BACKGROUND

The era of the Big Bang, and the first stars that formed in its immediate aftermath, lies tantalizingly just beyond the reach of conventional telescopes – but we can observe it in other ways. To radio telescopes the entire sky, in every direction, glows softly with a 'Cosmic Microwave Background' (CMB) radiation equivalent to the emission from an object at a temperature of just 2.7 °C (4.9 °F). This is the 'afterglow' of the Big Bang – radiation that began its journey in the cooling fireball itself, and has now been red-shifted to invisibility over 13.7 billion years of time.

■ MAPPING THE BACKGROUND

At first glance, the CMB radiation is entirely uniform in every direction, but cosmologists were sure it must contain some unevenness, since the Universe today is itself uneven and 'clumpy'. In reality, the temperature of the radiation fluctuates by a tiny amount – just one part in 100,000. The COBE satellite, launched in 1989, produced the first map of the CMB in 1992 (opposite), and the WMAP mission has since charted the fluctuations in far more detail (image at left). Together, these maps offer a unique view of the early variations in the Universe around which stars, galaxies, clusters and superclusters ultimately formed.

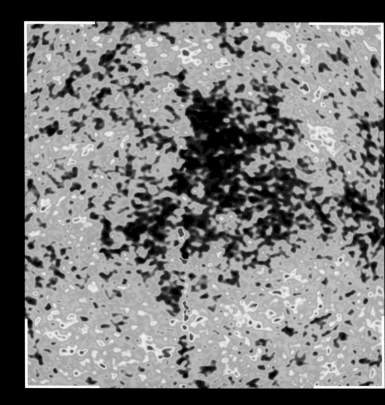

SOLAR SYSTEM DATA

Name	Distance from the Sun	Orbital period	Orbital eccentricity	Orbital inclination	Rotation period	Diameter	Mass	Density (water = 1)	Discovered
The Sun	-	-	-	-	24.6 Earth days	1,390,000 km (863,354 miles)	333,000 Earths	1.41	-
Mercury	57.91 million km (35.97 million miles)	87.97 days	0.21	7°	58.6 Earth days	4,880 km (3,032 miles)	0.055 Earths	5.43	-
Venus	108.2 million km (67.20 million miles)	224.7 days	0.01	3.4°	243 Earth days (retrograde)	12,104 km (7,518 miles)	0.816 Earths	5.24	-
Earth	149.6 million km (92.92 million miles)	365.26 days	0.02	0°	0.99 Earth days	12,756 km (7,922 miles)	5.97x10²¹ tonnes	5.52	-
Mars	227.9 million km (141.6 million miles)	686.98 days	0.09	1.9°	1.03 Earth days	6,794 km (4,220 miles)	0.108 Earths	3.93	-
Asteroid 1 Ceres	413.8 million km (257 million miles)	4.60 years	0.08	10.6°	0.37 Earth days	974 km (605 miles)	0.0002 Earths	2.1	1801
Jupiter	778.3 million km (483.4 million miles)	11.86 years	0.05	1.3°	0.41 Earth days	142,984 km (88,810 miles)	318.258 Earths	1.33	-
Saturn	1,429 million km (887.8 million miles)	29.46 years	0.06	2.5°	0.45 Earth days	120,536 km (74,868 miles)	95.142 Earths	0.69	-
Uranus	2,871 million km (1,783 million miles)	84.01 years	0.05	0.8°	0.72 Earth days (retrograde)	51,118 km (31,750 miles)	14.539 Earths	1.32	1781
Neptune	4,504 million km (2,798 million miles)	164.79 years	0.01	1.8°	0.67 Earth days	49,532 km (30,766 miles)	17.085 Earths	1.64	1846
Pluto	5,914 million km (3,673 million miles)	247.91 years	0.25	17.2°	6.39 Earth days (retrograde)	2,300 km (1,428 miles)	0.002 Earths	2.06	1930
Eris	10,120 million km (6,286 million miles)	557 years	0.44	44.1°	More than 0.33 Earth days	2,340 km (1,453 miles)	0.003 Earths	2.35	2005

SATELLITE DATA

Name	Orbits	Distance from parent planet	Orbital period	Orbital eccentricity	Orbital inclination	Rotation period	Diameter	Density (water = 1)	Discovered
Moon	Earth	384,400 km (238,800 miles)	27.32 days	0.05	5.1°	27.32 days	3,476 km (2,160 miles)	3.34	-
Phobos	Mars	9,000 km (5,590 miles)	0.32 days	0.02	1°	0.32 days	22 km (14 miles)	1.9	1877
Deimos	Mars	23,000 km (14,290 miles)	1.26 days	0	1.8°	1.26 days	12 km (8 miles)	1.8	1877
Metis	Jupiter	128,000 km (79,500 miles)	0.29 days	0	0°	?	45 km (28 miles)	2.8	1979
Adrastea	Jupiter	129,000 km (80,100 miles)	0.3 days	0	0°	?	16 km (10 miles)	4.5	1979
Amalthea	Jupiter	181,000 km (112,000 miles)	0.5 days	0	0.4°	0.5 days	250x146x128 km (155x91x79 miles)	1	1892
Thebe	Jupiter	222,000 km (138,000 miles)	0.67 days	0.02	0.8°	0.67 days	116x98x84 km (72x61x52 miles)	1.5	1979
Io	Jupiter	422,000 km (262,000 miles)	1.77 days	0	0°	1.77 days	3,642 km (2,262 miles)	3.53	1610
Europa	Jupiter	671,000 km (417,000 miles)	3.55 days	0.01	0.5°	3.55 days	3,130 km (1,944 miles)	2.99	1610
Ganymede	Jupiter	1.07 million km (665,000 miles)	7.15 days	0	0.2°	7.15 days	5,268 km (3,272 miles)	1.94	1610
Callisto	Jupiter	1.88 million km (1.17 million miles)	16.69 days	0.01	0.3°	16.69 days	4,806 km (2,986 miles)	1.85	1610
Leda	Jupiter	11.1 million km (6.89 million miles)	238.72 days	0.15	27°	?	16 km (10 miles)	2.7	1974
Himalia	Jupiter	11.5 million km (7.13 million miles)	250.57 days	0.16	28°	0.4 days	186 km (116 miles)	2.8	1904
Lysithea	Jupiter	11.7 million km (7.28 million miles)	259.22 days	0.11	29°	?	36 km (22 miles)	3.1	1938
Elara	Jupiter	11.7 million km (7.29 million miles)	259.65 days	0.21	28°	0.5 days	76 km (48 miles)	3.3	1905

SATELLITE DATA

Name	Orbits	Distance from parent planet	Orbital period	Orbital eccentricity	Orbital inclination	Rotation period	Diameter	Density (water = 1)	Discovered
Pasiphae	Jupiter	23.5 million km (14.60 million miles)	735 days (retrograde)	0.38	147°	?	50 km (31 miles)	2.9	1908
Sinope	Jupiter	23.7 million km (14.72 million miles)	758 days (retrograde)	0.28	153°	?	36 km (22 miles)	3.1	1914
Pan	Saturn	134,000 km (83,200 miles)	0.58 days	0	0°	?	28 km (17 miles)	?	1990
Atlas	Saturn	138,000 km (85,700 miles)	0.6 days	0	0°	?	30 km (18 miles)	?	1980
Prometheus	Saturn	139,000 km (86,300 miles)	0.61 days	0	0°	?	123x79x61 km (76x49x38 miles)	0.7	1980
Pandora	Saturn	142,000 km (88,200 miles)	0.63 days	0	0°	?	103x80x64 km (64x50x40 miles)	0.7	1980
Epimetheus	Saturn	151,000 km (93,800 miles)	0.69 days	0.01	0.3°	0.69 days	114 km (70 miles)	0.6	1980
Janus	Saturn	151,000 km (93,800 miles)	0.69 days	0.01	0.1°	0.69 days	195x194x156 km (121x120x97 miles)	0.65	1966
Mimas	Saturn	186,000 km (116,000 miles)	0.94 days	0.02	1.5°	0.94 days	398 km (248 miles)	1.14	1789
Enceladus	Saturn	238,000 km (148,000 miles)	1.37 days	0	0°	1.37 days	498 km (310 miles)	1.12	1789
Tethys	Saturn	295,000 km (183,000 miles)	1.89 days	0	1.1°	1.89 days	1,060 km (658 miles)	1	1684
Telesto	Saturn	295,000 km (183,000 miles)	1.89 days	0	0°	?	25 km (16 miles)	?	1980
Calypso	Saturn	295,000 km (183,000 miles)	1.89 days	0	0°	?	21 km (13 miles)	?	1980
Dione	Saturn	377,000 km (234,000 miles)	2.74 days	0	0°	2.74 days	1,120 km (696 miles)	1.44	1684
Helene	Saturn	377,000 km (234,000 miles)	2.74 days	0.01	0.2°	?	33 km (21 miles)	?	1980
Rhea	Saturn	527,000 km (327,000 miles)	4.52 days	0	0.4°	4.52 days	1,528 km (950 miles)	1.24	1672
Titan	Saturn	1.22 million km (759,000 miles)	15.95 days	0.03	0.3°	15.95 days	5,150 km (3,198 miles)	1.88	1655
Hyperion	Saturn	1.48 million km (920,000 miles)	21.28 days	0.1	0.4°	chaotic	328x260x214 km (204x161x133 miles)	1.4	1848
Iapetus	Saturn	3.56 million km (2.21 million miles)	79.33 days	0.03	14.7°	79.33 days	1,436 km (892 miles)	1.02	1671
Phoebe	Saturn	12.9 million km (8.04 million miles)	550.48 days (retrograde)	0.16	175.3°	115 days	230x220x210 km (143x137x130 miles)	0.06	1898
Cordelia	Uranus	50,000 km (31,100 miles)	0.34 days	0	0.1°	?	40 km (25 miles)	?	1986
Ophelia	Uranus	54,000 km (33,500 miles)	0.38 days	0	0.1°	?	43 km (27 miles)	?	1986

Name	Orbits	Distance from parent planet	Orbital period	Orbital eccentricity	Orbital inclination	Rotation period	Diameter
Bianca	Uranus	59,000 km (36,600 miles)	0.43 days	0	0.2°	?	51 km (32 miles)
Cressida	Uranus	62,000 km (38,500 miles)	0.46 days	0	0°	?	80 km (50 miles)
Desdemona	Uranus	63,000 km (39,100 miles)	0.47 days	0	0.2°	?	64 km (40 miles)
Juliet	Uranus	64,000 km (39,800 miles)	0.49 days	0	0.1°	?	94 km (58 miles)
Portia	Uranus	66,000 km (41,000 miles)	0.51 days	0	0.1°	?	135 km (84 miles)
Rosalind	Uranus	70,000 km (43,500 miles)	0.56 days	0	0.3°	?	72 km (45 miles)
Belinda	Uranus	75,000 km (46,600 miles)	0.62 days	0	0°	?	90 km (56 miles)
Puck	Uranus	86,000 km (53,400 miles)	0.76 days	0	0.3°	?	162 km (101 miles)
Miranda	Uranus	130,000 km (80,700 miles)	1.41 days	0	4.2°	1.41 days	472 km (294 miles)
Ariel	Uranus	191,000 km (119,000 miles)	2.52 days	0	0°	2.52 days	1,162 km (722 miles)
Umbriel	Uranus	266,000 km (165,000 miles)	4.14 days	0	0°	4.14 days	1,170 km (726 miles)
Titania	Uranus	436,000 km (271,000 miles)	8.71 days	0	0°	8.71 days	1,578 km (980 miles)
Oberon	Uranus	583,000 km (362,000 miles)	13.46 days	0	0°	13.46 days	1,522 km (946 miles)
Naiad	Neptune	48,000 km (29,800 miles)	0.29 days	0	0°	?	58 km (36 miles)
Thalassa	Neptune	50,000 km (31,100 miles)	0.31 days	0	4.5°	?	80 km (50 miles)
Despina	Neptune	53,000 km (32,900 miles)	0.33 days	0	0°	?	148 km (92 miles)
Galatea	Neptune	62,000 km (38,500 miles)	0.43 days	0	0°	?	158 km (98 miles)
Larissa	Neptune	74,000 km (46,000 miles)	0.55 days	0	0°	?	216x204x168 km (134x127x104 miles)
Proteus	Neptune	118,000 km (73,300 miles)	1.12 days	0	0°	?	436x416x402 km (271x258x250 miles)
Triton	Neptune	355,000 km (220,000 miles)	5.88 days (retrograde)	0	157°	5.88 days	2,706 km (1,680 miles)
Nereid	Neptune	5.51 million km (3.42 million miles)	360.13 days	0.75	29°	?	340 km (212 miles)
Charon	Pluto	20,000 km (12,400 miles)	6.39 days	0	98.8°	6.39 days	1,206 km (750 miles)
Nix	Pluto	49,000 km (30,400 miles)	24.86 days	0	0.2°	22.9	46 km (28 miles)
Hydra	Pluto	65,000 km (40,400 miles)	38.21 days	0.01	0.2°	23.4	60 km (38 miles)
Dysnomia	Eris	37,000 km (23,000 miles)	15.77 days	0.01	142°	?	50–125 km

GLOSSARY

Active galaxy

A galaxy that emits large amounts of energy from its central regions, probably generated as matter falls into a supermassive black hole at the heart of the galaxy.

Asteroid

One of the countless rocky worlds of the inner solar system, largely confined in the main Asteroid Belt beyond the orbit of Mars.

Astronomical unit

A unit of measurement widely used in astronomy, equivalent to Earth's average distance from the Sun – roughly 149.6 million km or 93 million miles.

Atmosphere

A shell of gases held around a planet or star by its gravity.

Barred spiral galaxy

A spiral galaxy in which the arms are linked to the hub by a straight bar of stars and other material.

Binary star

A pair of stars in orbit around one another. Because the stars in a binary pair were usually born at the same time, they allow a direct comparison of the way that stars with different properties evolve.

Black hole

A superdense point in space formed by a collapsing stellar core more than five times the mass of the Sun. A black hole's gravity is so powerful that even light cannot escape from it.

Brown dwarf

A so-called 'failed star' that never gains enough mass to begin the fusion of hydrogen in its core and start to shine properly. Instead, brown dwarfs radiate low-energy radiation (mostly infrared) through gravitational contraction and a more limited form of fusion.

Comet

A chunk of rock and ice from the outer reaches of the solar system. When comets fall into orbits that bring them close to the Sun, they heat up and their surface ices evaporate, forming a coma and a tail.

Dark nebula

A cloud of interstellar gas and dust that absorbs light, and only becomes visible when silhouetted against a field of stars or other nebulae.

Dwarf planet

Any object that is in an independent orbit around the Sun, and has sufficient gravity to pull itself into a roughly spherical shape, but which, unlike a true planet, has not cleared the region around it of other objects. Currently there are three known dwarf planets – the asteroid Ceres, and the Kuiper Belt Objects Pluto and Eris, but there are many more objects whose status is still uncertain.

Electromagnetic radiation

A form of energy consisting of combined electric and magnetic waves, able to propagate itself across a vacuum at the speed of light. The energy or temperature of an object emitting radiation affects its wavelength and other characteristics.

Elliptical galaxy

A galaxy consisting of stars in orbits that have no particular orientation, and generally lacking in star-forming gas. Ellipticals are among the smallest and largest galaxies known.

Emission nebula

A cloud of gas in space that glows at very specific wavelengths, producing a spectrum full of emission lines. These nebulae are usually energized by the high-energy light of nearby stars and are often associated with star-forming regions.

Flare

A huge release of superheated particles above the surface of a star, caused by a short-circuit in its magnetic field.

Fusion shell

A spherical shell of nuclear fusion spreading out through a star after it has exhausted a particular fuel supply in its core.

Galaxy

An independent system of stars, gas and other material with a size measured in thousands of light years.

Gamma rays

The highest-energy forms of electromagnetic radiation, with extremely short wavelengths, generated by the hottest objects and most energetic processes in the Universe.

Giant planet

A planet comprising a huge envelope of gas, liquid or slushy ice (various frozen chemicals), perhaps around a relatively small rocky core.

Globular cluster

A dense ball of ancient, long-lived stars, in orbit around a galaxy such as the Milky Way.

Helium fusion

Nuclear fusion of helium (formed by hydrogen fusion) into heavier elements (so-called metals). Most stars rely on helium fusion to keep on shining as they exhaust their supplies of hydrogen and near the end of their lives.

Hydrogen fusion

The nuclear fusion of hydrogen, the lightest element, into helium, the next lightest. Hydrogen fusion is the main power source for all stars for the majority of their lives, but it can proceed at different rates depending on conditions within a star.

Infrared

Electromagnetic radiation with slightly less energy than visible light. Infrared radiation is typically emitted by warm objects too cool to glow visibly.

Irregular galaxy
A galaxy with no obvious structure, generally rich in gas, dust, and star-forming regions.

Kuiper Belt
A doughnut-shaped ring of icy worlds directly beyond the orbit of Neptune. The largest known Kuiper Belt Objects are Pluto and Eris.

Light year
A common unit of astronomical measurement, equivalent to the distance travelled by light (or other electromagnetic radiation) in one year. A light year is equivalent to roughly 9.5 trillion km (5.9 million million miles).

Luminosity
A measure of the energy output of a star. Although luminosity is technically measured in watts, the stars are so luminous that it is simpler to compare them with the Sun. A star's visual luminosity (the energy it produces in visible light) is not necessarily equivalent to its overall luminosity in all radiations.

Main sequence
A term used to describe the longest phase in a star's life, during which it is relatively stable and shines by fusing hydrogen into helium at its core. During this period, the star obeys a general relationship that links its mass, size, luminosity and colour.

Multiple star
A system of two or more stars in orbit around one another (pairs of stars are also called binaries). Most of the stars in our galaxy are members of multiple systems rather than individuals like the Sun.

Nebula
Any cloud of gas or dust floating in space. Nebulae are the material from which stars are born, and into which they are scattered again at the end of their lives. The word means 'cloud' in Latin, and was originally applied to any fuzzy object in the sky, including some we now know to be star clusters or distant galaxies

Neutron star
The collapsed core of a supermassive star, left behind by a supernova explosion. A neutron star consists of compressed subatomic particles, and is the densest known object – though in the most massive stars, the core can collapse past the neutron star stage to form a black hole. Many neutron stars initially behave as pulsars.

Nova
A binary star system in which a white dwarf is pulling material from a companion star, building up a layer of gas around itself that then burns away in a violent nuclear explosion.

Nuclear fusion
The joining-together of light atomic nuclei (the central cores of atoms) to make heavier ones at very high temperatures and pressures, releasing excess energy in the process. Fusion is the process by which the stars shine.

Oort Cloud
A spherical shell of dormant comets, up to two light years across, surrounding the entire solar system.

Open cluster
A large group of bright young stars that have recently been born from the same star-forming nebula, and may still be embedded in its gas clouds.

Planet
A world that follows its own orbit around the Sun, is massive enough to pull itself into a spherical shape and which has cleared the space around it of other objects (apart from satellites). According to this definition, there are eight planets – Mercury, Venus, Earth, Mars, Jupiter, Saturn, Uranus and Neptune.

Planetary nebula
An expanding cloud of glowing gas sloughed off from the outer layers of a dying red giant star as it transforms into a white dwarf.

Pulsar
A rapidly spinning neutron star with an intense magnetic field that channels its radiation out along two narrow beams that sweep across the sky.

Radio
The lowest-energy form of electromagnetic radiation, with the longest wavelengths. Radio waves are emitted by cool gas clouds in space, but also by violent active galaxies and pulsars.

Red dwarf
A star with considerably less mass than the Sun – small, faint and with a low surface temperature. Red dwarfs fuse hydrogen into helium in their cores very slowly and live for much longer than the Sun, despite their size.

Red giant
A star passing through a phase of its life where its luminosity has increased hugely, causing its outer layers to expand and its surface to cool. Stars usually enter red giant phases when they exhaust the fuel supplies in their core.

Reflection nebula
A cloud of interstellar gas and dust that shines as it reflects or scatters light from nearby stars.

Rocky planet
A relatively small planet composed largely of rocks and minerals, perhaps surrounded by a thin envelope of gas and liquid.

Spectral lines

Dark or light bands in a spectrum of light that correspond to certain wavelengths. Bright emission lines can indicate that an object is emitting certain wavelengths, while dark bands silhouetted against a broad background spectrum indicate that something is absorbing the light on its way to us. In both cases, the location of the lines offers information on which atoms or molecules are involved.

Spectroscopic binary

A binary star that can only be detected thanks to the shifting of the lines in its spectrum as its two components swing around one another.

Spectrum

The spread-out band of light created by passing light through a prism or similar device. The prism bends light by different amounts depending on its wavelength and colour, so the spectrum reveals the precise intensities of light at different wavelengths.

Spiral galaxy

A galaxy consisting of a hub of old yellow stars, surrounded by a flattened disk of younger stars, gas and dust, with spiral arms marking regions of current star formation.

Star

A dense ball of gas that has collapsed into a spherical shape and become hot and dense enough at its centre to trigger nuclear fusion reactions that make it luminous.

Stellar wind

A stream of high-energy particles blasted off the surface of a star by the pressure of its radiation, that spread across the surrounding space.

Sun

The star at the centre of Earth's solar system. The Sun is a fairly average low-mass star, and a useful comparison for other stars. Its key properties include a diameter of 1.39 million km (860,000 miles), a mass of 2,000 trillion trillion tonnes, energy output of 300 trillion trillion watts and a surface temperature of 5,500°C (9,900°F).

Sun-like star

A yellow star with roughly the same mass, luminosity and surface temperature as the Sun. Stars like this are of particular interest to astronomers because they are long-lived, stable and any planets around them are potential havens for life.

Supergiant

A massive and extremely luminous star with between 10 and 70 times the mass of the Sun. Supergiants can have almost any colour, depending on how the balance of their energy output and their size affects their surface temperature.

Supermassive black hole

A black hole with the mass of millions of stars, believed to lie in the very centre of many galaxies. Supermassive black holes form from the collapse of huge gas clouds rather than the death of massive stars.

Supernova

A cataclysmic explosion marking the death of a star. Supernovae can be triggered when a heavyweight star exhausts the last of its fuel and its core collapses (forming either a neutron star or a black hole) or when a white dwarf in a nova system tips over its upper mass limit and collapses suddenly into a neutron star.

Supernova remnant

A cloud of superheated gas expanding from the site of a former supernova explosion.

Ultraviolet

Electromagnetic radiation with wavelengths slightly shorter than visible light, typically radiated by objects hotter than the Sun. The hottest stars give out much of their energy in the ultraviolet.

Variable star

A star that varies its brightness, either due to interaction with another star, or because of some feature of the star itself (most commonly a pulsation in size that may be periodic or irregular).

Visible light

Electromagnetic radiation with wavelengths between 400 and 700 nanometres (billionths of a metre), corresponding to the sensitivity of the human eye. Stars like the Sun emit most of their energy in the form of visible light.

White dwarf

A stellar remnant left behind by the death of a star with less than about eight times the Sun's mass. White dwarfs are the dense, slowly cooling cores of stars – typically very hot, but hard to see on account of their tiny size.

X-rays

High-energy electromagnetic radiation emitted by extremely hot objects and violent processes in the Universe. Material heated as it is pulled towards a black hole is one of the strongest sources of astronomical X-rays.

ACKNOWLEDGEMENTS

2–3: Credit for Hubble Image: NASA, ESA, N. Smith (University of California, Berkeley), and The Hubble Heritage Team (STScI/AURA). Credit for CTIO Image: N. Smith (University of California, Berkeley) and NOAO/AURA/NSF; 8–9: ESO/F. Kamphues; 10–11: ESO/S. Brunier; 12–13: NASA/JPL/University of Arizona; 15: Pikaia Imaging; 16: Pikaia Imaging; 18: NASA Goddard Space Flight Center Image by Reto Stöckli (land surface, shallow water, clouds). Enhancements by Robert Simmon (ocean color, compositing, 3D globes, animation). Data and technical support: MODIS Land Group; MODIS Science Data Support Team; MODIS Atmosphere Group; MODIS Ocean Group Additional data: USGS EROS Data Center (topography); USGS Terrestrial Remote Sensing Flagstaff Field Center (Antarctica); Defense Meteorological Satellite Program (city lights); 19: Jacques Descloitres, MODIS Land Rapid Response Team, NASA/GSFC; 20: European Space Agency. All rights reserved.; 21: Shutterstock/Sam Dcruz; 22: NASA / VRS / Science Photo Library; 23: Image by Robert Simmon, based on data © 2003 Geoeye. Caption by Holli Riebeek; 24: NASA; 25: NASA; 26: NASA; 27: Pikaia Imaging; 28: NASA; 29: NASA; 30: NASA/JPL/USGS; 31: NASA; 32: Courtesy of SOHO/EIT consortium. SOHO is a project of international cooperation between ESA and NASA; 33: TRACE; 34: Royal Swedish Academy of Sciences/Oddbjorn Engvold, Jun Elin Wiik, Luc Rouppe van der Voort, Institute of Theoretical Astrophysics of Oslo University; 35: TRACE; 36: Royal Swedish Academy of Sciences/Oddbjorn Engvold, Jun Elin Wiik, Luc Rouppe van der Voort, Institute of Theoretical Astrophysics of Oslo University; 37: Courtesy of SOHO/EIT consortium. SOHO is a project of international cooperation between ESA and NASA; 38: Image produced by NASA/Johns Hopkins University Applied Physics Laboratory/Arizona State University/Carnegie Institution of Washington. Image reproduced courtesy of Science/AAAS; 40: Image produced by NASA/Johns Hopkins University Applied Physics Laboratory/Arizona State University/Carnegie Institution of Washington. Image reproduced courtesy of Science/AAAS; 41: Image produced by NASA/Johns Hopkins University Applied Physics Laboratory/Arizona State University/Carnegie Institution of Washington. Image reproduced courtesy of Science/AAAS; 42: Image produced by NASA/Johns Hopkins University Applied Physics Laboratory/Arizona State University/Carnegie Institution of Washington. Image reproduced courtesy of Science/AAAS; 43: Image produced by NASA/Johns Hopkins University Applied Physics Laboratory/Arizona State University/Carnegie Institution of Washington. Image reproduced courtesy of Science/AAAS; 44: NASA/JPL; 45: NASA/JPL; 46: NASA/JPL; 47: NASA/JPL; 48: NASA/JPL/USGS; 49: NASA/JPL; 50: Viking Project, USGS, NASA; 51: ESA/DLR/FU Berlin (G. Neukum); 52: ESA/DLR/FU Berlin (G. Neukum); 53: NASA/JPL; 54: NASA/JPL-Caltech/MSSS; 56: NASA/JPL/Cornell; 57: NASA/JPL-Caltech/University of Arizona/Cornell/Ohio State University; 58: NASA/JPL/USGS; 59: ESA/DLR/FU Berlin (G. Neukum); 60: HiRISE, MRO, LPL (U. Arizona), NASA; 61: HiRISE, MRO, LPL (U. Arizona), NASA; 62: NASA/JPL/JHUAPL; 63: NASA/JPL/JHUAPL; 64: ESA 2010 MPS for OSIRIS Team MPS/UPD/LAM/IAA/RSSD/INTA/UPM/DASP/IDA; 65: ESA 2010 MPS for OSIRIS Team MPS/UPD/LAM/IAA/RSSD/INTA/UPM/DASP/IDA; 66: NASA/JPL; 67: NASA/JPL/University of Arizona; 68: NASA/JPL/Space Science Institute; 69: NASA/JPL; 70: NASA, ESA, A. Simon-Miller (Goddard Space Flight Center), N. Chanover (New Mexico State University), and G. Orton (Jet Propulsion Laboratory); 71: NASA, ESA, I. de Pater and M. Wong (University of California, Berkeley); 72: Pikaia Imaging; 73: NASA/JPL/USGS; 74: NASA/JPL/University of Arizona; 75: NASA/JPL/University of Arizona; 76: NASA/JPL/DLR; 77: NASA/JPL/University of Arizona; 78: NASA/JPL/DLR; 79: NASA/JPL/University of Arizona; 80: NASA/JPL; 81: NASA/JPL/Brown University; 82: NASA/JPL/Brown University; 83: NASA/JPL; 84: NASA/JPL/DLR; 85: NASA/JPL; 86: NASA/JPL/Space Science Institute; 87: NASA/JPL/Space Science Institute; 88: NASA/JPL/Space Science Institute; 89: Cassini Imaging Team, SSI, JPL, ESA, NASA; 90: NASA/JPL/Space Science Institute; 91: NASA/JPL/Space Science Institute; 92: NASA/JPL/Space Science Institute; 93: NASA/JPL/Space Science Institute; 94: NASA/JPL/Space Science Institute; 95: NASA/JPL/Space Science Institute; 96: NASA/JPL/Space Science Institute/Universities Space Research Association/Lunar & Planetary Institute; 97: NASA/JPL/Space Science Institute; 98: NASA/JPL/Space Science Institute; 99: NASA/JPL/Space Science Institute; 100: NASA/JPL/Space Science Institute; 101: NASA/JPL/Space Science Institute; 102: NASA/JPL/Space Science Institute; 103: NASA/JPL/Space Science Institute; 104: NASA/JPL/Space Science Institute/Universities Space Research Association/Lunar & Planetary Institute; 105: NASA/JPL/Space Science Institute; 106: NASA/JPL/Space Science Institute; 107: NASA/JPL/University of Arizona; 108: ESA/NASA/JPL/University of Arizona; 109: ESA/NASA/JPL/University of Arizona; 110: NASA/JPL/University of Arizona/DLR; 111: NASA/JPL; 112: NASA/JPL/University of Arizona/Ames/Space Science Institute; 113: NASA/JPL/Space Science Institute; 114: NASA/JPL/Space Science Institute; 115: NASA/JPL/Space Science Institute; 116: NASA/JPL/Space Science Institute; 117: NASA/JPL/Space Science Institute; 118: NASA/JPL/Space Science Institute; 119: NASA/JPL/Space Science Institute; 120: NASA/JPL/Space Science Institute; 121: Lawrence Sromovsky, University of Wisconsin-Madison/ W. M. Keck Observatory; 122: NASA/JPL; 123: N ASA/JPL/STScI; 124: NASA/JPL; 125: NASA/JPL; 126: NASA/JPL; 127: NASA/JPL; 128: NASA/JPL; 129: NASA/JPL; 130: NASA/JPL; 131: NASA/JPL; 132: NASA/JPL; 133: NASA/JPL; 134: NASA, ESA, and M. Buie (Southwest Research Institute); 135: NASA, ESA, H. Weaver (JHUAPL), A. Stern (SwRI), and the HST Pluto Companion Search Team; 136: Harvard College Observatory/Science Photo Library; 137: A-/MPE, 1986, 1996; 138: NASA, ESA, and A. Schaller (for STScI); 139: NASA, ESA, and M. Brown (California Institute of Technology); 140–1: NASA, ESA, C.R. O'Dell (Vanderbilt University), M. Meixner and P. McCullough (STScI); 142–3: Pikaia Imaging; 144: Pikaia Imaging; 146: K. Noll (Hubble Heritage PI/STScI), C. Luginbuhl (USNO), F. Hamilton (Hubble Heritage/STScI); 147: ESO/L Emerson/VISTA. Acknowledgment: Cambridge Astronomical Survey Unit; 148: Credit for Hubble Image: NASA, ESA, N. Smith (University of California, Berkeley), and The Hubble Heritage Team (STScI/AURA), Credit for CTIO Image: N. Smith (University of California, Berkeley) and NOAO/AURA/NSF; 149: Credit for Hubble Image: NASA, ESA, N. Smith (University of California, Berkeley), and The Hubble Heritage Team (STScI/AURA), Credit for CTIO Image: N. Smith (University of California, Berkeley), and The Hubble Heritage Team (STScI/AURA), Credit for CTIO Image: N. Smith (University of California, Berkeley) and NOAO/AURA/NSF; 150: Credit for Hubble Image: NASA, ESA, N. Smith (University of California, Berkeley), and The Hubble Heritage Team (STScI/AURA), Credit for CTIO Image: N. Smith (University of California, Berkeley) and NOAO/AURA/NSF; 151: ESO ; 152: Nathan Smith, University of Minnesota/NOAO/AURA/NSF; 153: NASA, The Hubble Heritage Team (AURA/STScI); 154: NASA, ESA, T. Megeath (University of Toledo) and M. Robberto (STScI); 155: ESO/L. Emerson/VISTA. Acknowledgment: Cambridge Astronomical Survey Unit; 156: NASA,ESA, M. Robberto (Space Telescope Science Institute/ESA) and the Hubble Space Telescope Orion Treasury Project Team; 157: NASA,ESA, M. Robberto (Space Telescope Science Institute/ESA) and the Hubble Space Telescope Orion Treasury Project Team; 158: NASA, The NICMOS Group (STScI, ESA) and The NICMOS Science Team (Univ. of Arizona); 159: NASA, H. Ford (JHU), G. Illingworth (UCSC/LO), M. Clampin (STScI), G. Hartig (STScI), the ACS Science Team, and ESA; 160: ESO; 161: NASA, ESA, STScI, J. Hester and P. Scowen (Arizona State University); 162: NASA, ESA, and The Hubble Heritage Team (STScI/AURA); 163: NASA, ESA, and The Hubble Heritage Team (STScI/AURA); 164: T.A.Rector (NRAO/AUI/NSF and NOAO/AURA/NSF) and B.A.Wolpa (NOAO/AURA/NSF); 165: NASA, ESA, STScI, J. Hester and P. Scowen (Arizona State University); 166: A. Caulet (ST-ECF, ESA) and NASA; 167: ESO; 168: T.A. Rector/University of Alaska Anchorage and WIYN/AURA/NSF; 169: NASA, ESA, and The Hubble Heritage Team (STScI/AURA), Acknowledgment: P. McCullough (STScI); 170: NASA, ESA and L. Ricci (ESO); 171: ESO/L Emerson/VISTA & R. Gendler. Acknowledgment: Cambridge Astronomical Survey Unit. ; 172: Capella Observatory/Josef Pöpsel, Beate Behle, Dr Stefan Binnewies; 173: J. Morse/STScI, and NASA; 174: NASA and C.R. O'Dell (Vanderbilt University); 175: NASA and C.R. O'Dell (Vanderbilt University); 176: NASA, C.R. O'Dell and S.K. Wong (Rice University); 177: NASA; K.L. Luhman (Harvard-Smithsonian Center for Astrophysics, Cambridge, Mass.); and G. Schneider, E. Young, G. Rieke, A. Cotera, H. Chen, M. Rieke, R. Thompson (Steward Observatory, University of Arizona, Tucson, Ariz.); 178: NASA and The Hubble Heritage Team (STScI/AURA); 179: NASA, ESA and AURA/Caltech; 180: NASA, ESA, and G. Meylan (Ecole Polytechnique Federale de Lausanne); 181: ESO; 182: NASA, ESA, and the Hubble SM4 ERO Team; 183: ESO; 184: ESO/IDA/Danish 1.5 m/ R. Gendler, U.G. Jørgensen, J. Skottfelt, K. Harpsøe; 185: ASA, ESA and Jesús Maz Apellyniz (Instituto de astrofisica de Andaluca, Spain). Acknowledgement: Davide De Martin (ESA/Hubble); 186: NASA, ESA, P. Kalas, J. Graham, E. Chiang, E. Kite (Univ. California, Berkeley), M. Clampin (NASA/Goddard), M. Fitzgerald (Lawrence Livermore NL), K. Stapelfeldt, J. Krist (NASA/JPL); 187: NASA, ESA, P. Kalas, J. Graham, E. Chiang, E. Kite (Univ. California, Berkeley), M. Clampin (NASA/Goddard), M. Fitzgerald (Lawrence Livermore NL), K. Stapelfeldt, J. Krist (NASA/JPL); 188: ESO; 189: A. Fujii; 190: A. Fujii; 191: ESO and P. Kervella; 192: Nathan Smith (University of California, Berkeley), and NASA; 193: ESO; 194: NASA, ESA, and the Hubble SM4 ERO Team; 195: Margarita Karovska (Harvard-Smithsonian Center for Astrophysics) and NASA; 196: NASA, ESA and H.E. Bond (STScI); 197: NASA, ESA and H.E. Bond (STScI); 198: J.P. Harrington and K.J. Borkowski (University of Maryland), and NASA; 199: NASA/JPL-Caltech/J. Hora (Harvard-Smithsonian CfA); 200: R. Corradi (Isaac Newton Group) and D. Goncalves (Inst. Astrofisica de Canarias); 201: NASA, ESA, HEIC, and The Hubble Heritage Team (STScI/AURA). Acknowledgment: R. Corradi (Isaac Newton Group of Telescopes, Spain) and Z. Tsvetanov (NASA); 202: H. Bond et al., Hubble Heritage Team (STScI /AURA), NASA; 203: NASA/JPL-Caltech/Harvard-Smithsonian CfA ; 204: NASA, ESA, C.R. O'Dell (Vanderbilt University), M. Meixner and P. McCullough (STScI); 205: NASA, NOAO, ESA, the Hubble Helix Nebula Team, M. Meixner (STScI), and T.A. Rector (NRAO); 206: NASA, NOAO, ESA, the Hubble Helix Nebula Team, M. Meixner (STScI), and T.A. Rector (NRAO); 207: NASA/JPL-Caltech/Univ of Ariz.; 208: NASA, ESA, and the Hubble SM4 ERO Team; 209: NASA, ESA and A.Zijlstra (UMIST, Manchester, UK); 210: A. Fujii; 211: NASA, H.E. Bond and E. Nelan (Space Telescope Science Institute, Baltimore, Md.); M. Barstow and M. Burleigh (University of Leicester, U.K.); and J.B. Holberg (University of Arizona); 212: NASA and The Hubble Heritage Team (STScI/AURA); 213: NASA, ESA, J. Hester and A. Loll (Arizona State University); 214: NASA/CXC/ASU/ J. Hester et al.;

Quercus Publishing Plc
21 Bloomsbury Square
London
WC1A 2NS

First published in 2011

A catalogue record of this book is available from the British Library

ISBN: 978 0 85738 345 7

Printed and bound in China

10 9 8 7 6 5 4 3 2 1